Thinking Spatially Using GIS

Eileen J. Napoleon
Erin A. Brook

Student Workbook **Level 1**

ISBN 978-1-58948-184-8 $12.95
107038 RRD4M04/08jc
Printed in the USA

Thinking Spatially Using GIS

Eileen J. Napoleon
Erin A. Brook

Student Workbook

ESRI Press
REDLANDS, CALIFORNIA

ESRI Press, 380 New York Street, Redlands, California 92373-8100

12 11 10 09 08   1 2 3 4 5 6 7 8 9 10

Printed in the United States of America

Ask for ESRI Press titles at your local bookstore or order by calling 1-800-447-9778. You can also shop online at www.esri.com/esripress. Outside the United States, contact your local ESRI distributor.

ESRI Press titles are distributed to the trade by the following:

In North America:
Ingram Publisher Services
Toll-free telephone: (800) 648-3104
Toll-free fax: (800) 838-1149
E-mail: customerservice@ingrampublisherservices.com

In the United Kingdom, Europe, and the Middle East:
Transatlantic Publishers Group Ltd.
Telephone: 44 20 7373 2515
Fax: 44 20 7244 1018
E-mail: richard@tpgltd.co.uk

Contents

Name: ______________________________

Date: ______________________________

Module 1, Lesson 1

Magellan crosses the Atlantic Ocean

M1
L1

After Christopher Columbus found the New World in 1492, Spain and Portugal were eager to conquer and claim new lands. The two world powers decided to divide the world in half by drawing a line that ran through the Atlantic Ocean. Based on this line, Spain could claim new lands in the western half of the world, and Portugal could claim lands in the eastern half.

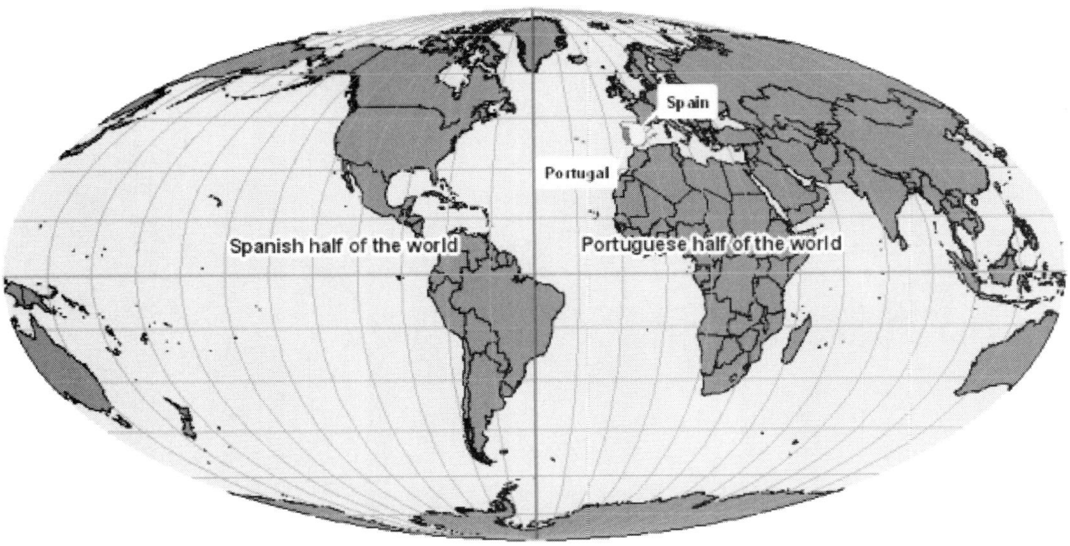

This shows the approximate location of the dividing line between the Spanish (western) half of the world and the Portuguese (eastern) half of the world.

Both countries wanted to claim the Spice Islands because of the cloves and other valuable spices that could be found there. But it was unclear where the Spice Islands were in relation to the dividing line because the islands were on the far side of the world from Europe.

Ferdinand Magellan, a Portuguese explorer born in 1480, led the first expedition to sail completely around the world. We call this a *circumnavigation*. Like Christopher Columbus, Magellan believed he could get to the Spice Islands by sailing west. Magellan knew that he would have to sail around or through the Americas to do it.

Portugal turned down Magellan's plan, but he won support from the king of Spain. Magellan thought he could find a new route to the Spice Islands and perhaps claim those lands for Spain. In 1519, Magellan set sail with five ships and 270 men.

Let's follow Magellan's voyage and find out the new route he discovered.

Step 1: Start AEJEE.

1. Ask your teacher how to start the AEJEE software.
2. Click the **Maximize** button at the top right of the window. If you are not sure how to do this, ask your teacher.

   Now the AEJEE window fills your screen.

Step 2: Open the project.

1. Click the **Open** button. You see the Open window.
2. Choose **OurWorld1** and click **Open**.

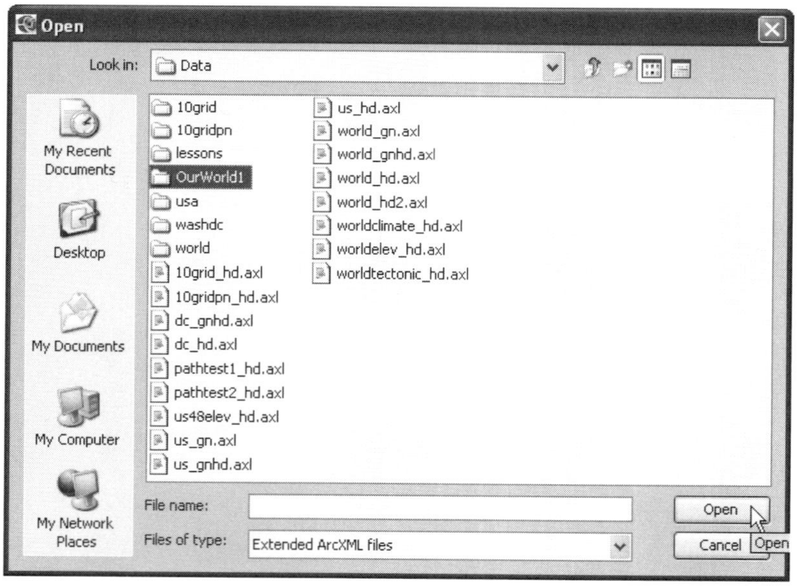

3. Choose **Module1** and click **Open**.

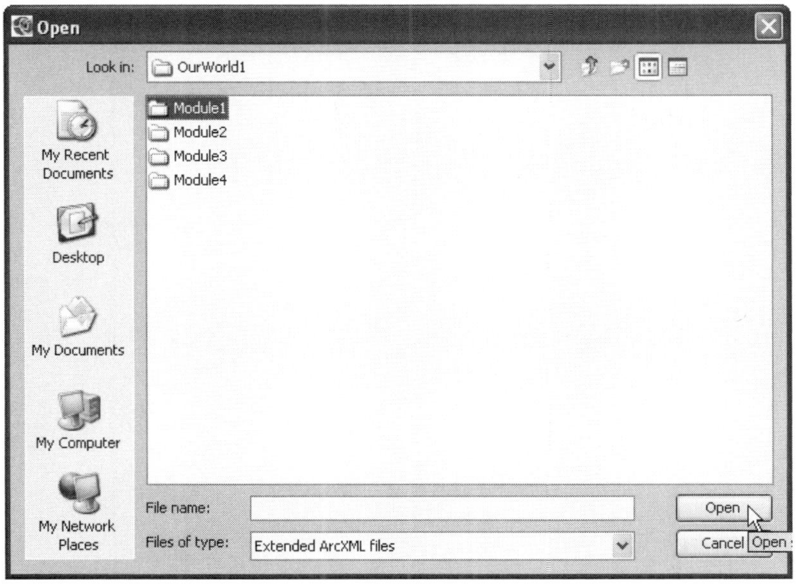

M1
L1

4. Choose **MagellanAtlantic.axl** and click **Open**.

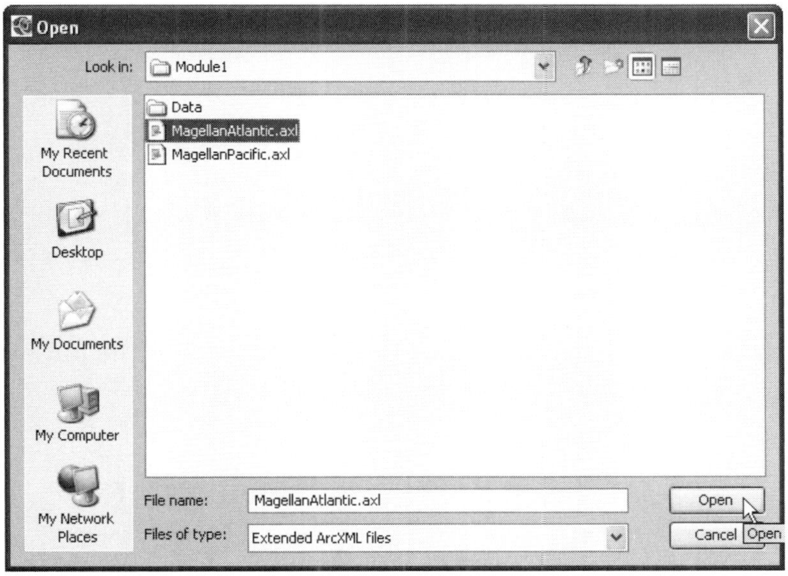

A map of the world appears on your screen.

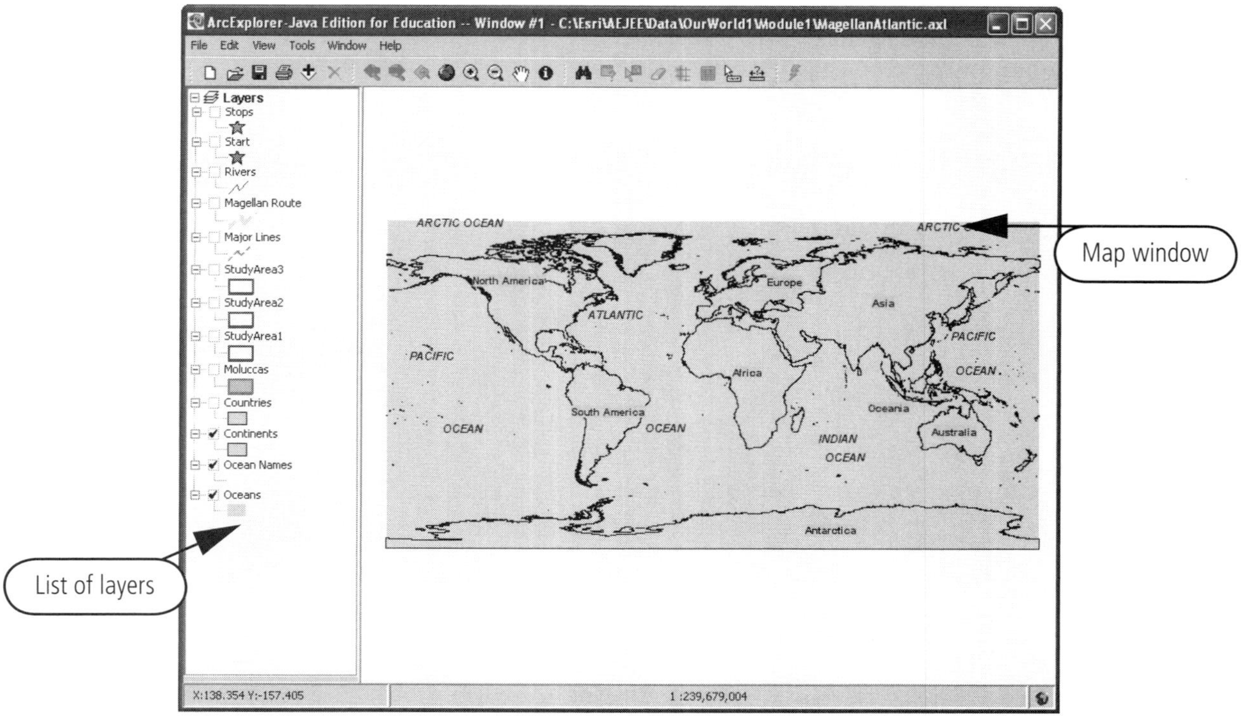

On the left side you see a list of layers. Layers are used to show geographic data on GIS maps. Each layer has a name and a legend. You can turn each layer on and off. At the top of the window, you see menus and buttons that you will use during this activity.

Q1 **Look at the continents on the map and fill in the missing letters in the following statements.**

a. Europe is touching the continent of __ __ __ A.

b. The __ __ L __ __ __ I __ Ocean separates the continent of __ F __ __ __ A from South America.

c. The __ __ C __ __ __ C Ocean separates the continent of North America from __ S __ __.

Step 3: Locate the Spice Islands, also known as the Moluccas.

1. Turn on the Moluccas layer by clicking the box next to Moluccas.

2. Look on your map for a small red area in the Far East.

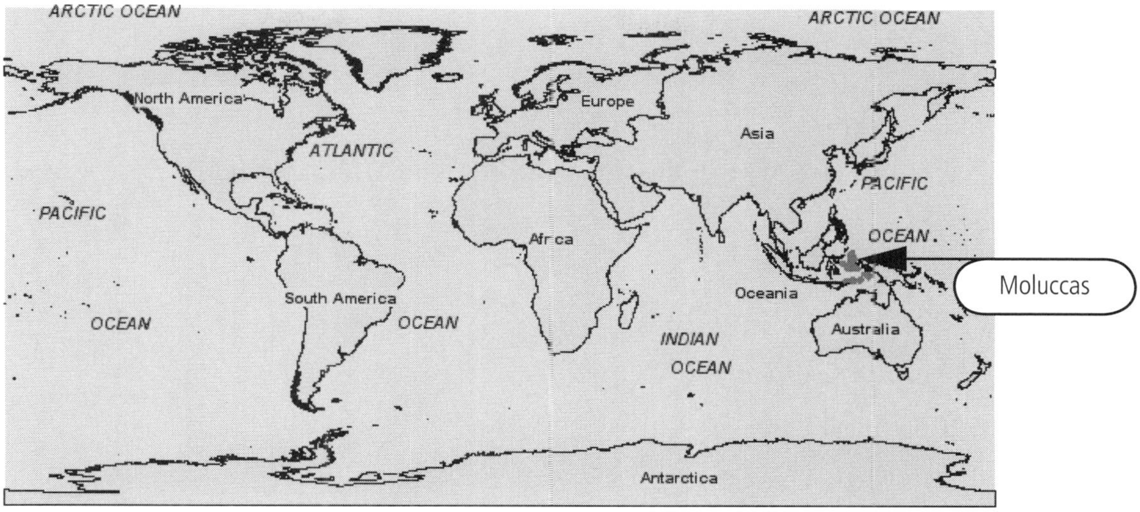

M1
L1

3. Click the **Continents** layer name to highlight it. It becomes active.

> **Note:** Making a layer active tells the GIS that this is the layer you want to work with or learn about.

4. Click the **Identify** tool. It allows you to get information about the active layer.
5. Click the **Moluccas** on the map.

The Identify Results window opens. On the left you see the name of the continent you clicked. On the right you see additional information about that continent.

Q2 **What is the name of the continent where the Moluccas are found?**

a. Australia

b. Oceania

c. Asia

6. Close the **Identify Results** window.

7. Turn on the **Major Lines** layer by clicking the box next to its name. It shows the major lines of latitude and longitude (e.g., the equator and the prime meridian).

Q3 **What major line of latitude passes through the Moluccas? (Circle the correct answer.)**

a. Tropic of Cancer

b. Equator

c. Tropic of Capricorn

d. Prime meridian

8. Turn off the **Major Lines** layer by clicking the check mark next to its name.

Step 4: View Magellan's route.

You'll turn on different layers to see where Magellan sailed.

1. Turn on the **Countries** layer by clicking the box next to Countries in the list of layers. The outlines of all the countries appear on the map.

2. Turn off the **Continents** layer by clicking the check mark next to Continents.

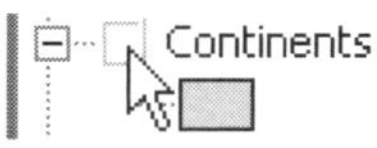

3. Turn on the **Start** layer by clicking the box next to its name. A red star appears showing where Magellan started his voyage.

4. Turn on the **Magellan Route** layer by clicking the box next to its name.

The route Magellan followed around the world appears on the map.

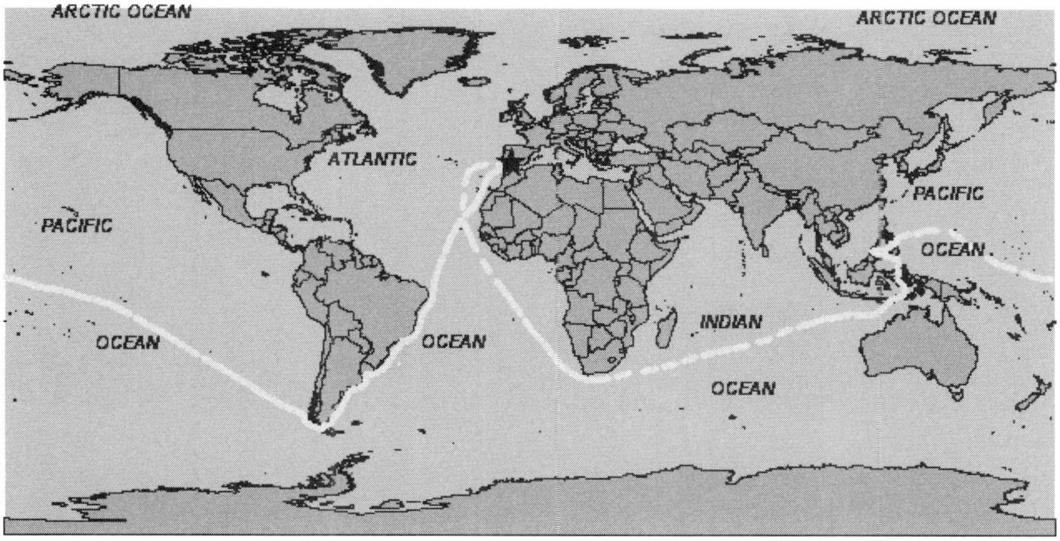

M1
L1

5. Click the **Start** layer name to highlight it. It becomes active.

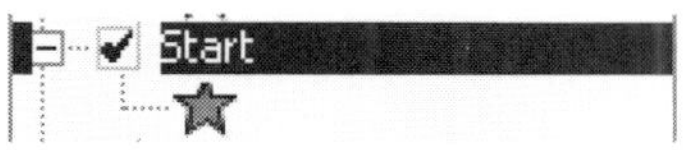

6. Click the **Identify** tool.
7. Click the red star on the map. The Identify Results window opens.

Q4 **What is the name of the city where Magellan began his voyage?** ______________

Q5 **What country is this city in? (Hint: Look on the right side of the Identify Results window.)** ______________

8. Close the **Identify Results** window.
9. Turn on the **Major Lines** layer in the list of layers.

Q6 **Magellan started at the red star and traveled west. How many times did he cross the equator to return to the starting point? (Circle the correct answer.)**

a. **One time**

b. **Two times**

c. **Three times**

d. **Four times**

A true circumnavigation requires crossing the equator at least once.

Q7 **Did Magellan spend more time north of the equator or south of the equator? (Circle the correct answer.)**

a. North of the equator

b. South of the equator

The equator divides the world in half between north and south. Each half is called a *hemisphere*.

Step 5: Follow Magellan as he starts his voyage in September of 1519 and passes through two groups of islands.

You'll zoom in for a closer look at Magellan's route.

1. Turn off the **Moluccas** layer by clicking the box next to its name.

2. Turn on the **StudyArea1** layer.
3. Click the **StudyArea1** layer name to highlight it. It becomes active.

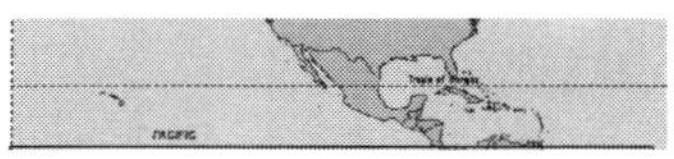

4. Click the **Zoom to Active Layer** button.

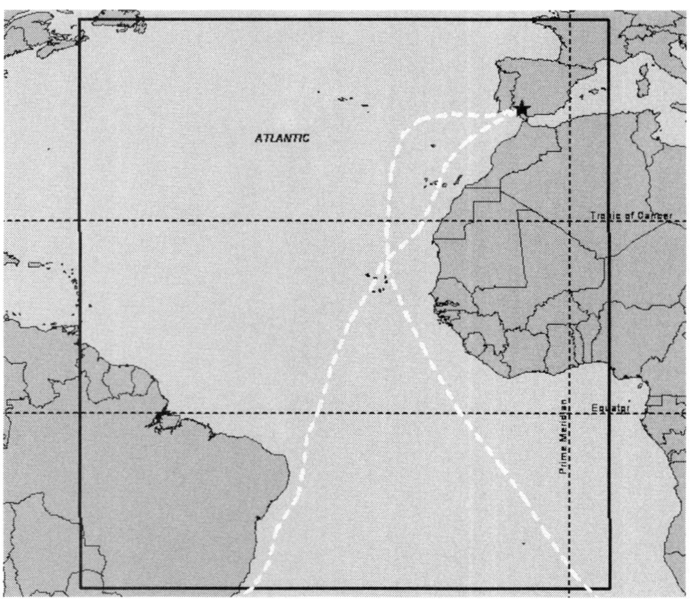

Q8 **How did your map change?**

a. **The map shows more detail (larger scale).**

b. **The map shows less detail (smaller scale).**

5. Turn off the **StudyArea1** layer.
6. Turn on the **Stops** layer.

Magellan passed through two groups of islands on his way to the equator.

Q9 **What direction did Magellan have to sail to reach these islands? (Circle the correct answer.)**

a. **Northeast**

b. **North**

c. **Southeast**

d. **Southwest**

7. Click the **Stops** layer name to highlight it. It becomes active.

8. Click the **Identify** tool.
9. Click the first stop Magellan reached after leaving the start point.

Q10 **What is the name of the group of islands at this stop? ________________Islands**

10. Close the **Identify Results** window.
11. Click the next stop.

Q11 **What is the name of the group of islands at this stop? ________________Islands**

12. Close the **Identify Results** window.

Step 6: Magellan crossed the equator in October of 1519, one month after starting his voyage.

Near the end of October, Magellan's fleet neared the equator. The ships entered an area of calm water with no winds. This area is called the *doldrums*. The doldrums are located between 5 degrees north and 5 degrees south of the equator (see the orange band in the picture below). Magellan had to sail out of the doldrums in order to reach the *trade winds* south of the equator. Trade winds are warm, steady winds that blow all the time. They are located between 5 and 30 degrees north and south of the equator.

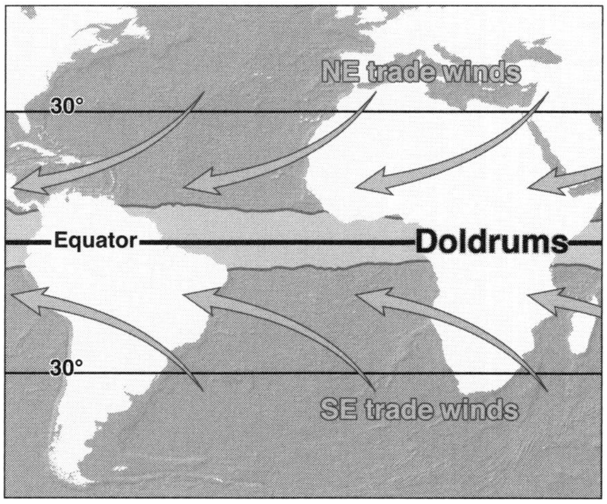

The *doldrums* are a zone of still air close to the equator. The *trade winds* are a zone of steady winds that blow from east to west.

In the doldrums, sailing ships often stood still, sometimes for days, making it frustrating for sailors. The weather there is hot and sticky. Only sudden thunderstorms with strong winds allowed ships to move.

Q12 **Based on what you know so far, why did it take Magellan three weeks to cross the equator and reach the trade winds?**

____

____

____

Q13 **Look at the picture above. After crossing the equator, Magellan was trying to reach Brazil on the east coast of South America. Did the trade winds blow Magellan toward Brazil or away from Brazil?**

a. Toward Brazil

b. Away from Brazil

Step 7: Magellan sailed along the coast of Brazil looking for a headland that would provide passage across the South American continent.

Magellan was told about a great headland near the southern tip of Brazil. (A headland is a place where the land is shaped like a "head" and sticks out into the ocean.) Magellan thought this headland could be a passage. He wanted to find it.

You'll find Brazil and select it on the map.

1. Click the **Find** button. The Find window opens.
2. In the Value box, type **Brazil** (make sure you type it correctly).
3. In the **Layers to Search** list, click **Countries**.
4. Click **Find**.

   The information appears on the right side of the window.

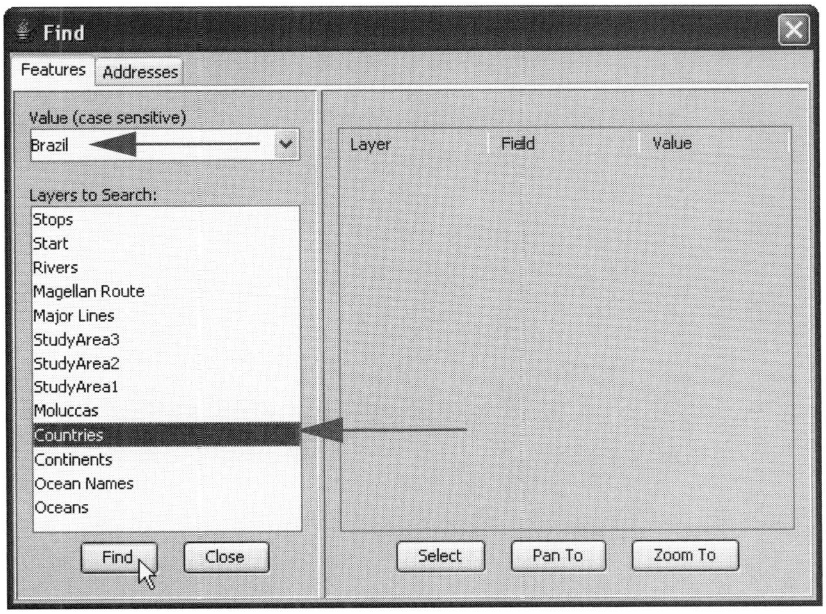

5. Click **Select** at the bottom of the window.
6. Close the **Find** window.

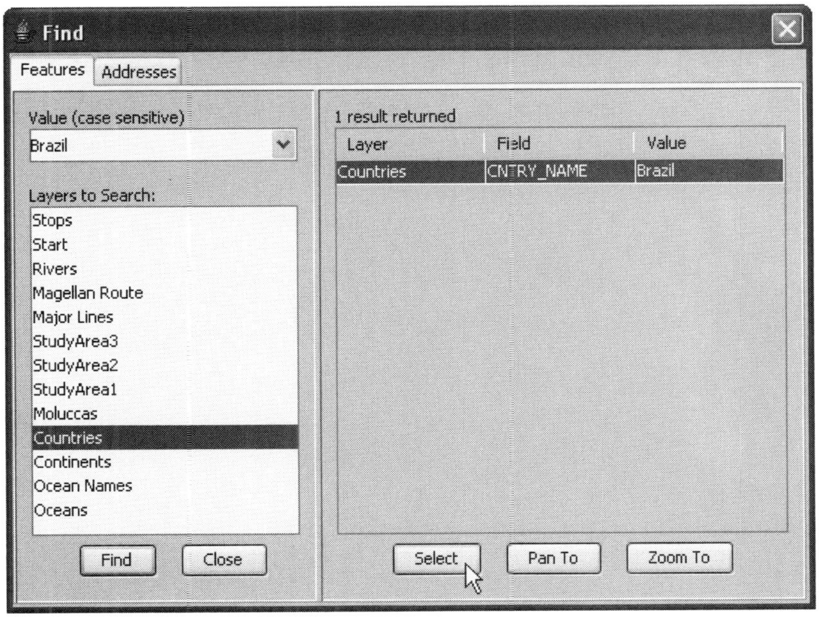

The country of Brazil is now highlighted in yellow on your map. You'll zoom out a little so you can see all of it.

7. Click the **Zoom Out** tool.

8. Place your mouse on the spot where Magellan's route to Brazil crosses the equator.

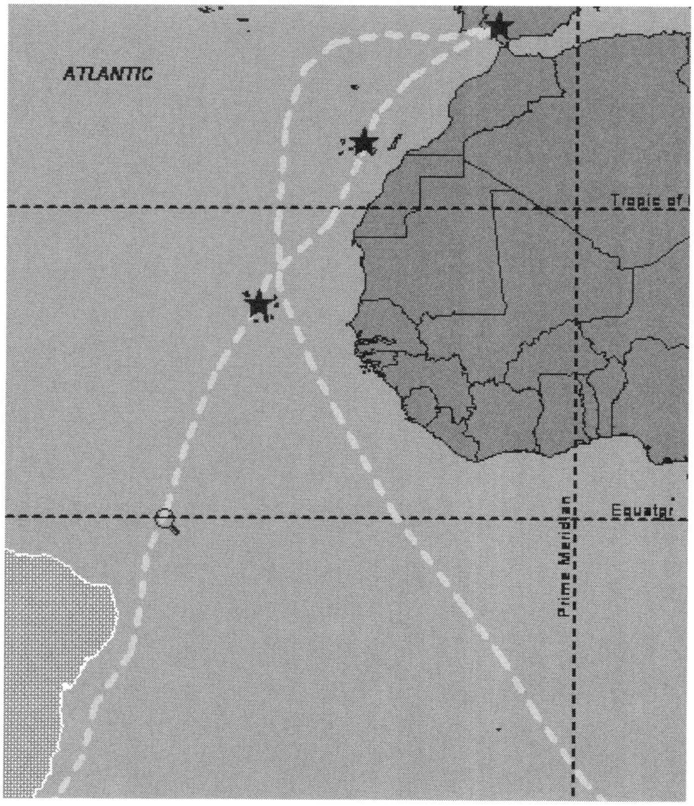

9. Click one time.

Q14 **How did your map change? (Circle all the correct answers.)**

a. The map shows less detail (smaller scale).

b. The map shows more detail (larger scale).

c. The map shows a bigger area.

d. The map shows a smaller area.

> **Note:** If you click in the wrong place, click the Previous Extent button and try again.

10. Click the **Stops** layer name to highlight it.

11. Click the **Identify** tool.

(Q15) What is the name of the place in Brazil where Magellan landed?

____

Magellan stopped here to make repairs and stock up on food. How far did Magellan travel to get here? You will use the Measure tool to find out.

12. Close the **Identify Results** window.

13. Click the **Measure** tool and choose **Nautical Miles** from the list.

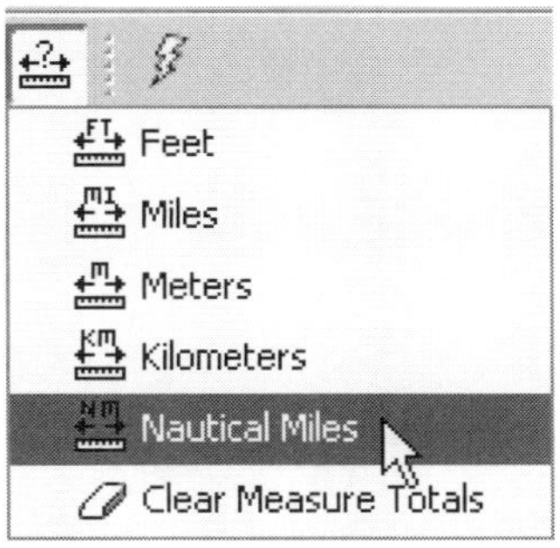

Nautical Miles are slightly longer than regular miles.

- A regular mile is 5,280 feet in length.
- A nautical mile is 6,076.1 feet in length.

14. Click and *hold* your mouse on the starting place of Magellan's voyage.
15. Drag your mouse to Magellan's stop in Brazil. The gray box in the top left corner of your map shows the length of the line you are drawing.
16. Release the mouse button.

(Q16) Look at the gray box. How many nautical miles did Magellan travel? (Hint: Use the number next to Total.) __, __ __ __.__ __ __ nm

Step 8: Magellan reached the headland on Christmas Day, 1519, three months after starting his voyage.

Magellan and his fleet left Brazil to sail along the east coast of South America, looking for the headland that he was told about. Just south of Brazil, they found it.

Before continuing, you will label the Stops.

1. Right-click **Stops** and choose **Properties**. This opens the Properties window.

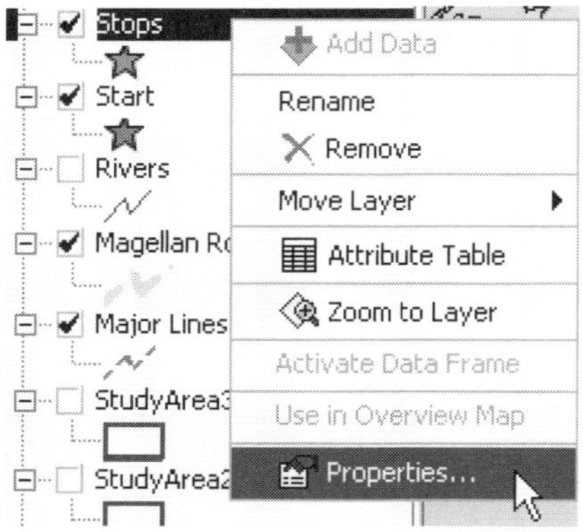

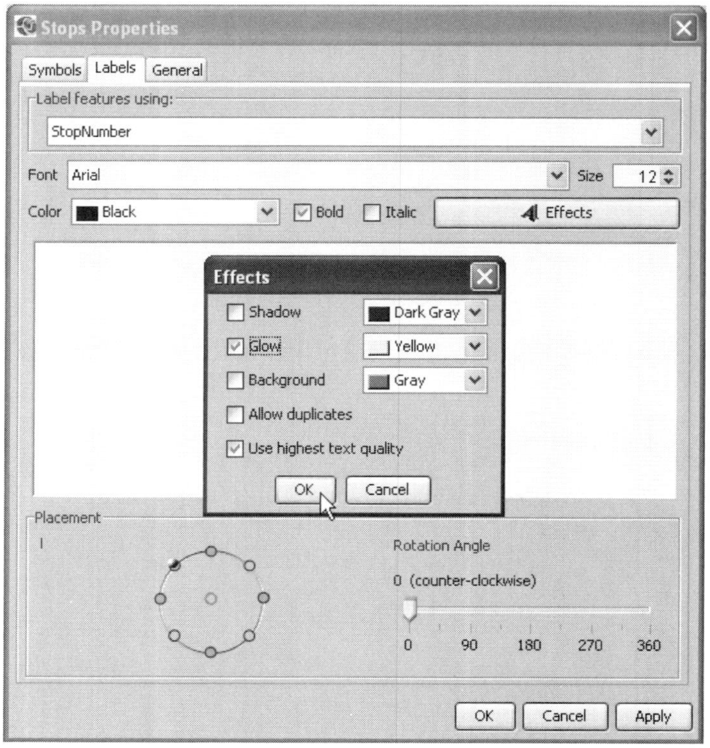

2. Click the **Labels** tab at the top.
3. Under **Label features using**, choose **StopNumber**.
4. Increase the **Size** to **12**.
5. Click the box next to **Bold** to check it.
6. Click the **Effects** button.
7. Click the box next to **Glow**.
8. Click **OK**.
9. Click **OK** again.

Now the stops are labeled with numbers that have a yellow glow around them.

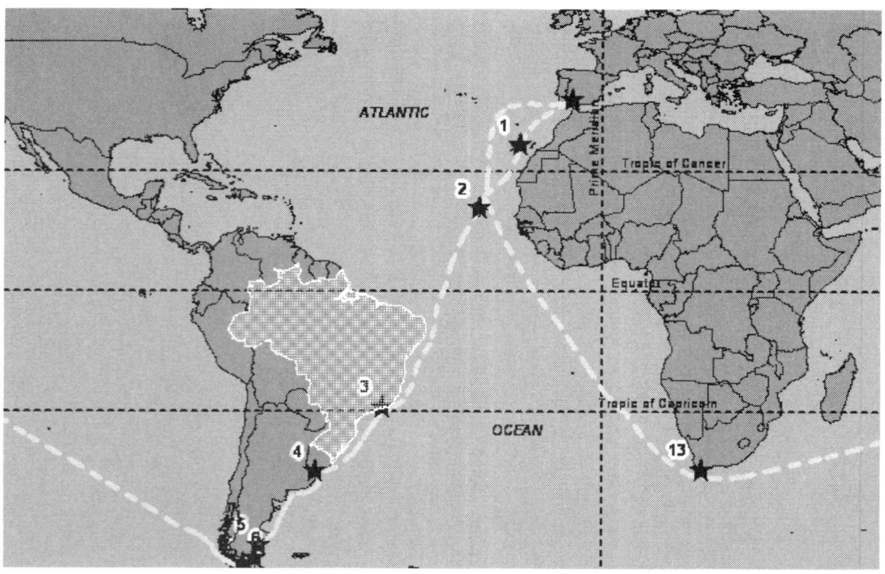

10. Look at the map and find Stop 4.

Magellan sent one of his ships to explore this location. He hoped it was a place where ships could cross to the other side of the South American continent.

11. Turn on the **Rivers** layer by clicking the box next to its name.
12. Click the **Rivers** layer name to highlight it. It becomes active.

13. Click the **Identify** tool.
14. Click the river that leads into the place where Magellan explored.

Q17 **What is the name of the river where Magellan's crew explored?**

____

15. Close the **Identify Results** window.

Q18 **Why do you think Magellan couldn't sail his ships to the other side of the continent from this place?**

____

____

____

Magellan left the river in early February of 1520. His crew wanted to head north for the winter, but Magellan decided to sail south toward the tip of South America.

To see the tip of South America, you will pan the map.

16. Click the **View** menu and choose **Pan Panel**.

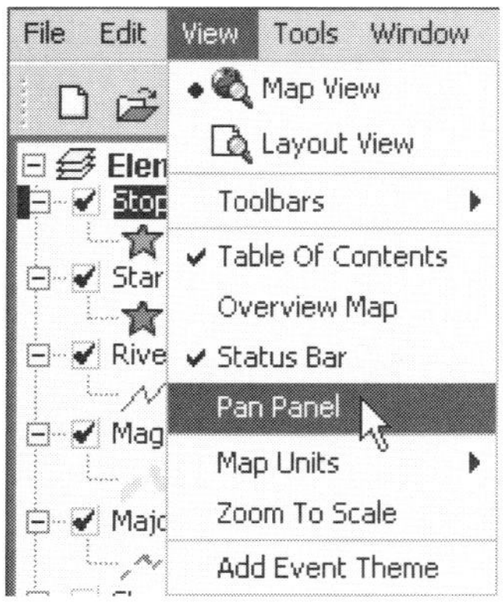

You see a frame around your map. Each side of the frame has a small white arrow. By clicking one of these arrows, you can move or pan the map up (north), down (south), left (west), or right (east).

17. To pan south, click once on the white arrow at the bottom of the map.

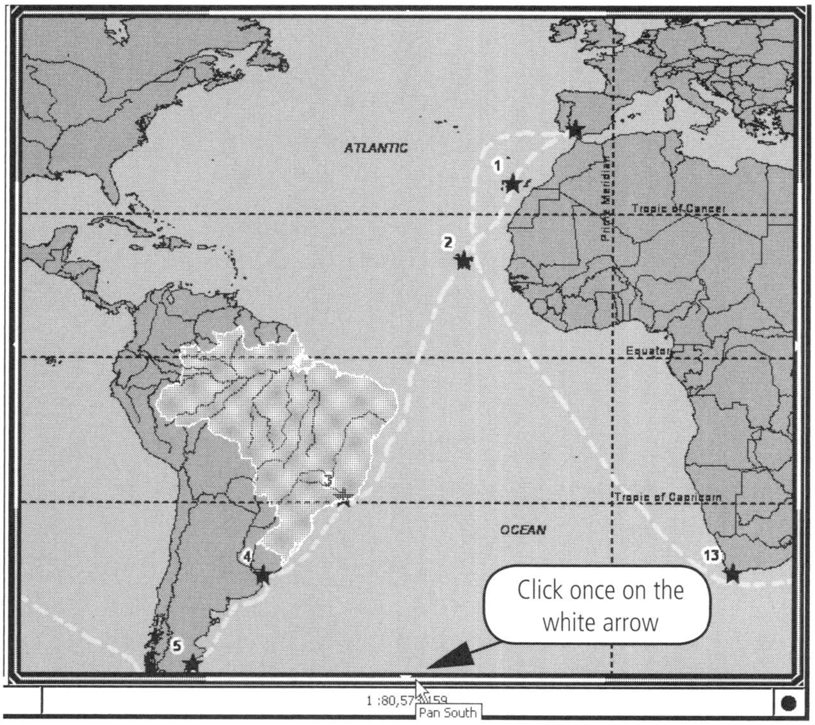

Q19 **What happened to your map? Describe the changes in your own words.**

____

____

____

Step 9: Follow Magellan's route south along a deserted coast.

Magellan's ships faced strong winds and big waves as they sailed for about two months along the deserted coast. Bad weather seriously damaged three of Magellan's ships. By late March 1520, Magellan was forced to stop and spend the winter. He found a safe harbor at Stop 5.

Q20 What major line of latitude is south of the southern tip of South America? (Circle the correct answer.)

a. Equator

b. Tropic of Capricorn

c. Arctic Circle

d. Antarctic Circle

Q21 As Magellan sailed south to Stop 5, do you think the temperature got warmer or colder?

a. Colder

b. Warmer

1. Turn on **StudyArea2** by clicking the box next to its name.
2. Click the **StudyArea2** layer name to highlight it. It becomes active.

3. Click the **Zoom to Active Layer** button.

Q22 Look at your map. Do you see more detail or less detail?

a. More detail

b. Less detail

4. Turn off **StudyArea2**.
5. Click the **Stops** layer name to highlight it.
6. Click the **Identify** tool.
7. Click Stop 5 (the star).

Q23 **What is the name of Stop 5? ______________________________**

Magellan faced mutiny at this stop. A mutiny is when a group, like the crew on a ship, refuses to obey the person in charge, in this case, Magellan. Fortunately for Magellan, the mutiny failed, and he punished the leaders.

8. Close the **Identify Results** window.

Step 10: Magellan discovered an ocean.

In October 1520, more than a year after he began his voyage, Magellan and his crew found the headland they were looking for. They called it the Cape of the Eleven Thousand Virgins. This is Stop 6.

Q24 **Click Stop 6 to find out what this stop is called today. (Hint: Use the Identify tool.)**

____

Two ships were blown west of Stop 6. Magellan thought they were lost. He found them in a deep, narrow channel of water (called a strait) that ran westward through the mountains. He followed this channel all the way across the tip of the South American continent.

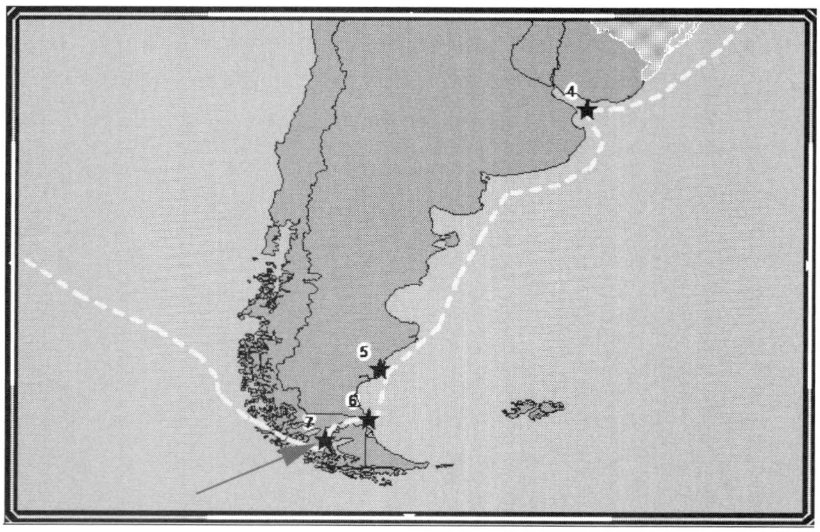

Q25 **Click Stop 7 to find out what this famous passage is called today. (Hint: Use the Identify tool.)**

____

1. Close the **Identify Results** window.
2. Turn on the **StudyArea3** layer. You can only see part of this layer now.
3. Click the **StudyArea3** layer name to highlight it. It becomes active.

4. Click the **Zoom to Active Layer** button.

The scale of the map changed, so you can see a bigger area.

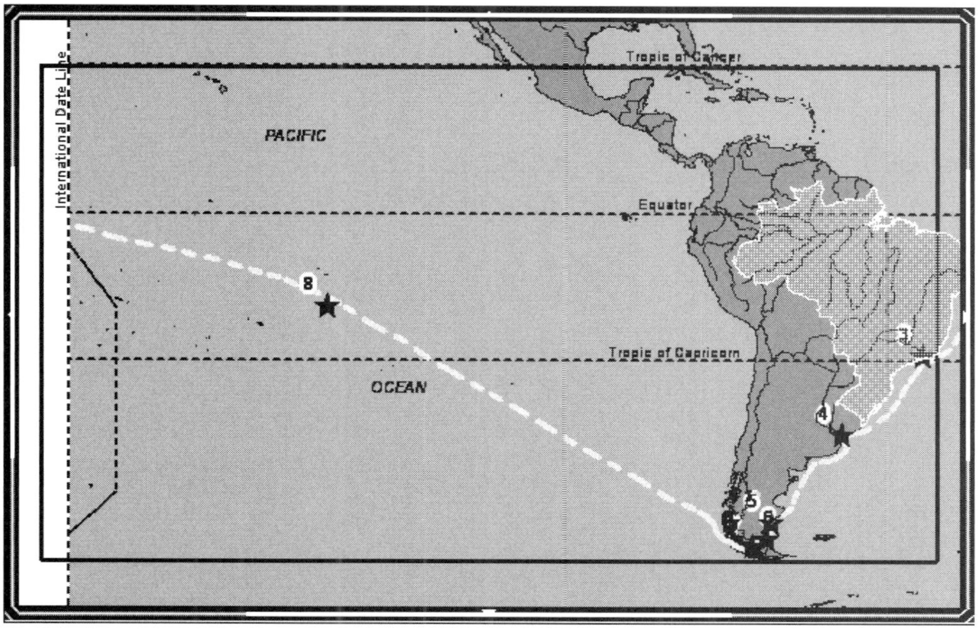

5. Turn off **StudyArea3**.

Q26 **What is the name of the ocean that Magellan reached on the west side of South America? (Circle the correct answer.)**

a. Atlantic Ocean

b. Pacific Ocean

Step 11: Save your work and exit AEJEE.

1. Ask your teacher how and where to save your work.

M1 L1

2. Click the **File** menu and choose **Exit**.

Q27 **Write a short paragraph explaining what Magellan accomplished to this point. (Hint: What bodies of water and what major lines of latitude did he cross? What continents and countries did he visited?)**

____

____

____

Conclusion

At this point, Magellan and his crew were the first Europeans to cross the South American continent and reach the ocean on the other side. They still needed to cross a big ocean to reach the continent of Asia and find the Spice Islands. With GIS, you have followed his path, gotten information about his stops, and measured distances along the way. In the next lesson, you will continue to follow Magellan's exploration.

Name: ____________________

Date: ____________________

Module 1, Lesson 2

Magellan crosses the Pacific Ocean

Ferdinand Magellan was the first European explorer to reach the Pacific Ocean when his expedition sailed through an opening, or strait, near the tip of South America in 1520. He named the ocean Mar Pacifico, which means peaceful sea. The strait, which connected the Atlantic and Pacific oceans, was later named for him.

At that point in his journey, Magellan and his fleet had been at sea for more than a year. He had lost two of his five ships. Now he would cross the Pacific Ocean with three ships, looking for the coast of Asia and the Spice Islands. However, he had no idea the Pacific Ocean would be so big!

Step 1: Start AEJEE.

1. Ask your teacher how to start the AEJEE software.
2. Click the **Maximize** button at the top of the window. Now the AEJEE window fills your screen.

Step 2: Open the project.

1. Click the **Open** button.
2. Navigate to your **OurWorld1\Module1** folder.
3. Choose **MagellanPacific.axl**.
4. Click **Open**.

   When the project opens, you see a map of the world showing Magellan's route on your screen.

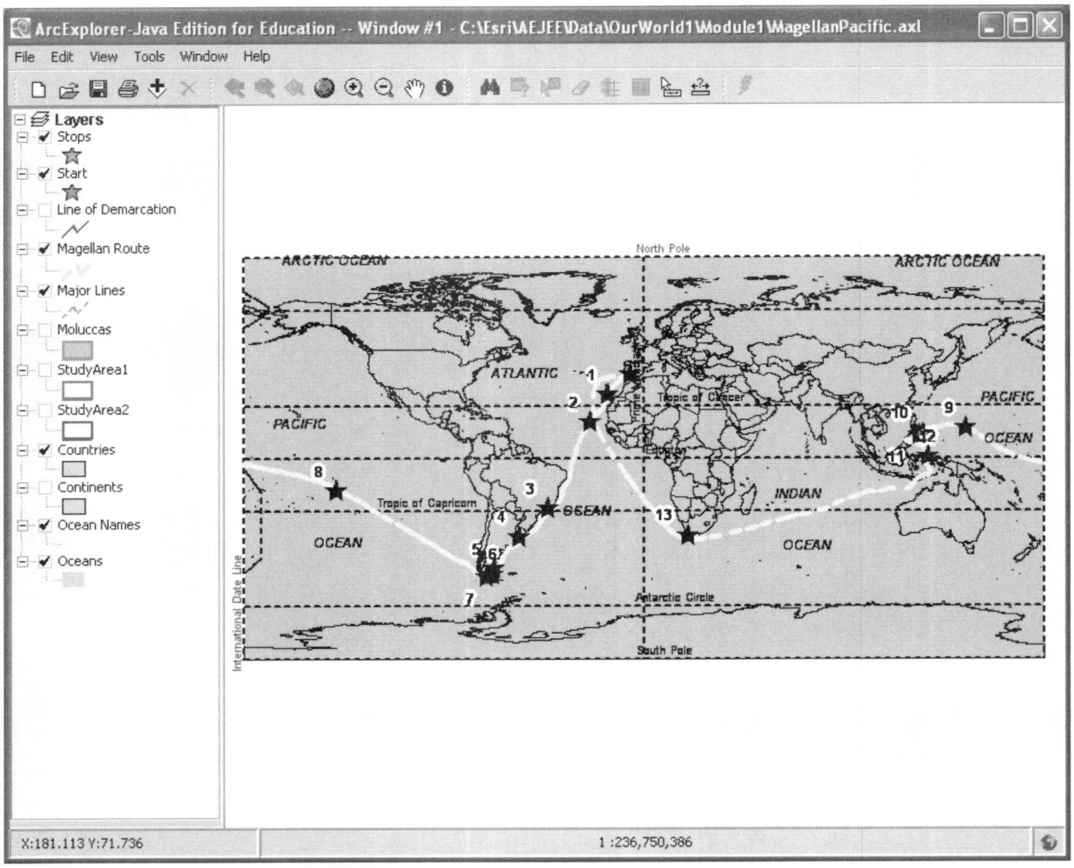

Notice the list of layers on the left side. Layers are used to show geographic data on GIS maps. Each layer has a name and a legend. You can turn each layer on and off. At the top of the window, you see menus and buttons that you will use during this activity.

Step 3: Follow Magellan on his Pacific voyage.

You are going to focus on the area where Magellan began to cross the Pacific Ocean.

1. Turn on the **StudyArea1** layer by clicking the box next to its name.
2. Click the **StudyArea1** layer name to highlight it. It becomes active.

3. Click the **Zoom to Active Layer** button.

The map redraws, and you are zoomed in to the Pacific Ocean.

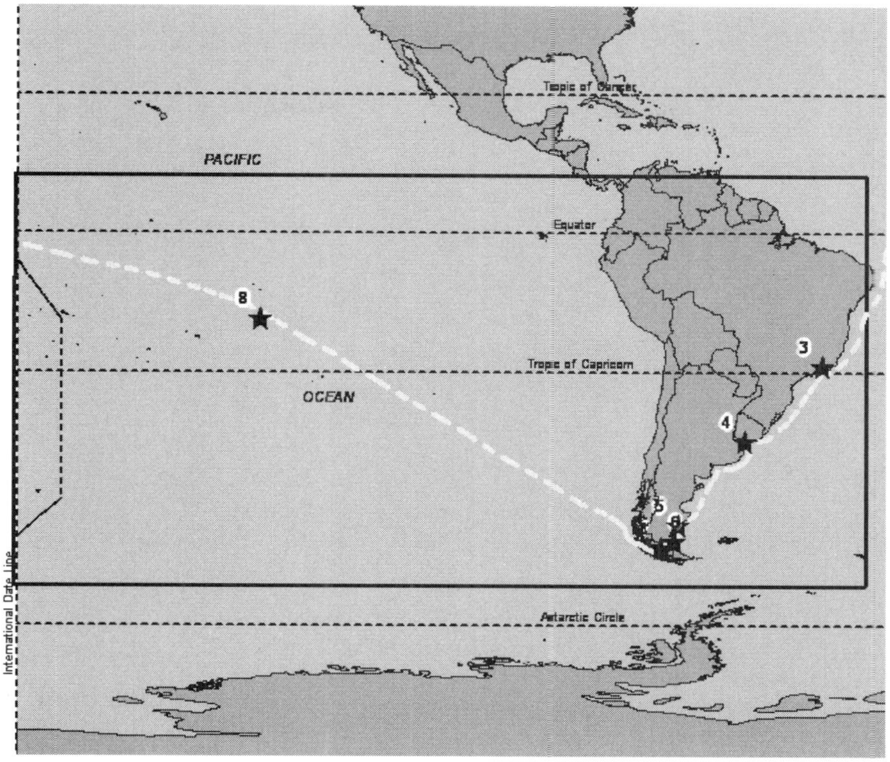

4. Turn off the **StudyArea1** layer by clicking the check mark next to its name.

Magellan hoped to find the coast of Asia, but the ships sailed for weeks without spotting land. Their food supplies rotted and began to disappear. After six weeks, his men began to die of scurvy, a disease caused by not having enough vitamin C. By mid-January 1521, more than a third of his men were too weak to walk, and they had only enough water for one sip a day.

Finally, on January 25, they spotted a small island at approximately 139 degrees west longitude and 14 degrees south latitude.

5. Place your mouse on the map while you look in the bottom left corner of the AEJEE window.

The numbers you see represent longitude (X) and latitude (Y). Longitude and latitude are used to identify the exact location of any place on the earth.

The X represents longitude—east or west of the prime meridian (a minus sign means west). The Y represents latitude—north or south of the equator (a minus sign means south).

6. Move your mouse until you find **X:-139 Y:-14**.

Q1 **What stop number is at this location?**______________________________

Q2 **Is this location east or west of the prime meridian?**

a. East

b. West

Q3 **Is this location north or south of the equator?**

a. North

b. South

You can see the stop numbers, but you can't see their names. You will turn on MapTips so you can see the name of each stop when you place your mouse on it.

7. Click the **MapTips** button. The MapTips window opens.
8. In the **Layers** list, scroll down and click **Stops**.
9. Under **Fields**, click **StopName**.
10. Click **Set MapTips**.
11. Click **OK**.

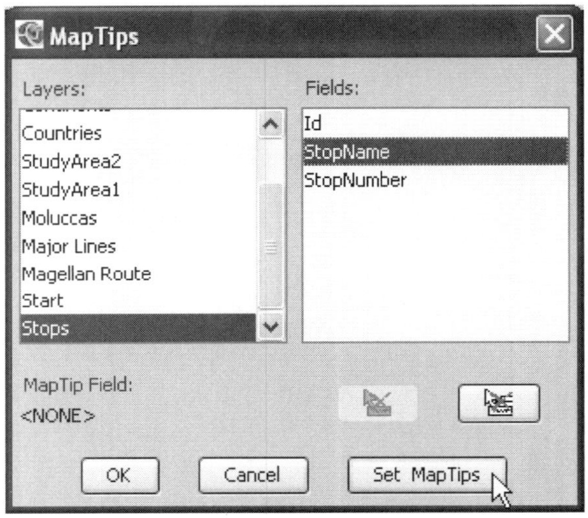

12. Hold your mouse over the red star at Stop 8 (don't click).

The name of the stop appears in a small box called a MapTip.

> **Note:** If you don't see the MapTip, move your mouse away from the red star, then move it back again.

Q4 **What is the name of the island at this stop?** _ _ _ _ _ _ _ _

Step 4: They left the island on January 28, 1521, and sailed onward.

To follow Magellan's route across the Pacific, you will change the shape of the map.

1. Click the **Tools** menu and choose **Projection** from the list.

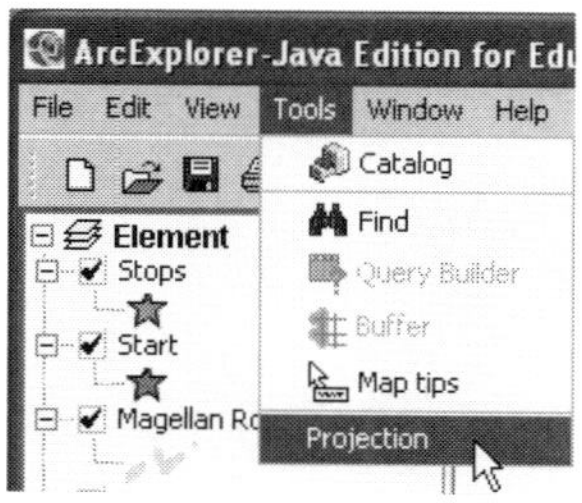

2. In the **Select Coordinate System** window, click the **Custom** tab.
3. Click the drop-down arrow next to **Projection**.
4. Scroll down in the list until you see **Orthographic**, and click it.

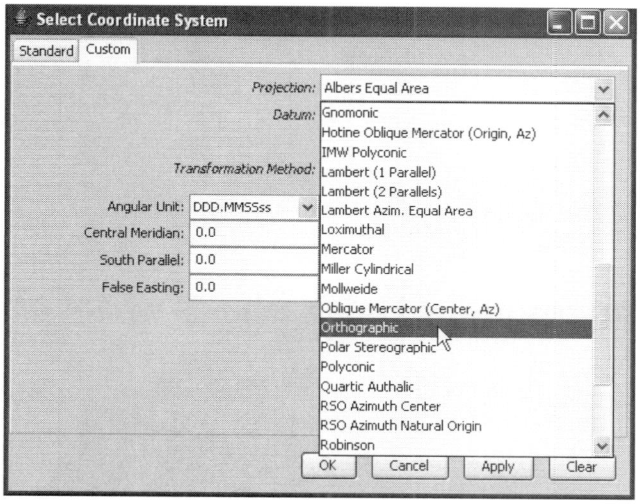

5. Click the **Center Longitude** box, and change the value from 0.0 to **-160.0**. (Be sure to include the minus sign!)

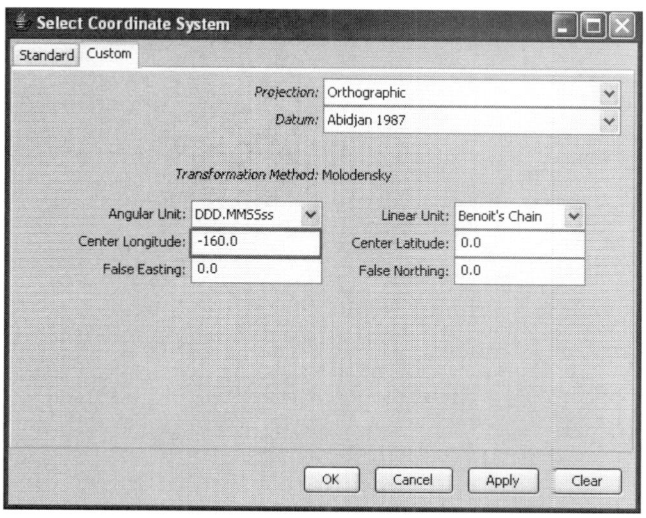

M1
L2

6. Click **OK**.

7. Click the **Zoom to Full Extent** button.

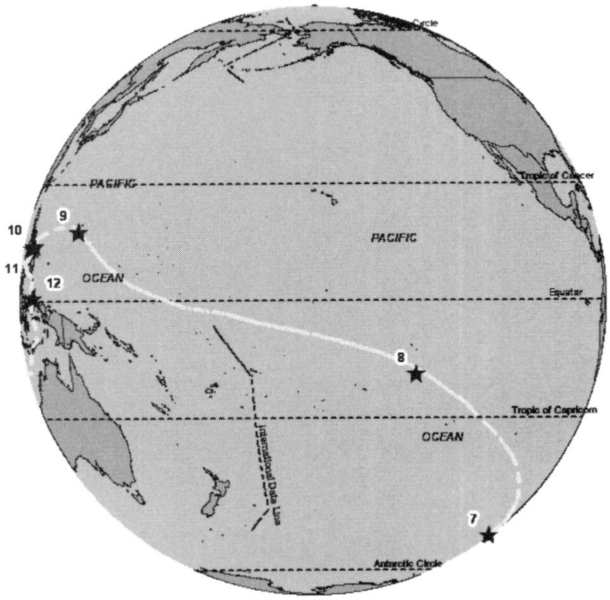

Q5 **What happened to your map? Describe the changes in your own words.**

____

____

Step 5: Magellan's route missed most of the Pacific islands where he could have found food.

When Magellan left Stop 8, he decided to head west and slightly north. By March 4, about five weeks after leaving the island at Stop 8, Magellan's ship ran out of food, leaving only a dozen men strong enough to do any work.

On March 5, just when things seemed hopeless, they spotted land at Stop 9.

Q6 **What is the name of this stop? (Hint: Hold your mouse over the stop and read the MapTip.)**____

At this stop, native people carrying clubs, spears, and shields surrounded Magellan's ships. Magellan used crossbows, a type of bow and arrow, to kill some of them. These native people, called Chamorros, left in their canoes. Magellan stole their water and food and used cannons to bomb their village. Magellan's fleet sailed away as hundreds of Chamorros returned in their canoes to fight.

Before you turn the globe again, you will measure the distance Magellan traveled across the Pacific Ocean.

1. Click the **Measure** tool and choose **Nautical Miles**.

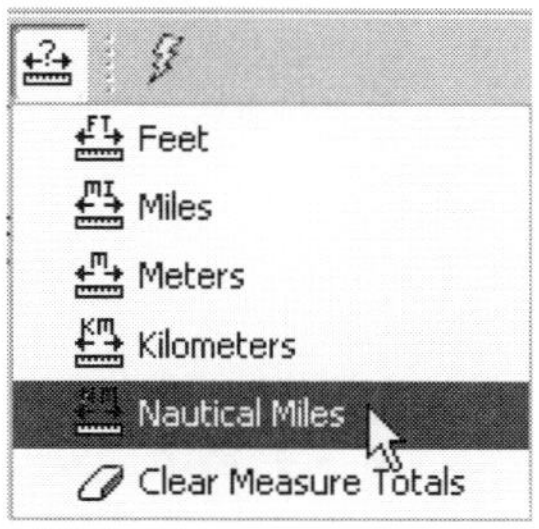

2. Click and *hold* down your mouse button on Stop 7.
3. Drag your mouse to Stop 10 or 11 (they are on top of each other).
4. Let go of your mouse button.

> **Note:** If you make a mistake, click the **Measure** tool and choose **Clear Measure Totals**. Then try again.

You see three lines on the map. Each shows a different measurement. The three lines are explained in the box below.

> When you measure on a projected map in AEJEE, you see three different lines:
> - The red line is a straight line on the screen. It looks like the shortest distance between two points, but because the earth is round, it may not be the shortest distance in real life.
> - The blue line is a great circle route. It is the shortest distance between two points if you used a piece of string on a real globe to measure the distance.
> - The pink line is a rhumb line. This line follows the same compass direction for the entire trip.

Q7 **Look at the curved blue line on your map and the blue (geodesic) measurement in the gray box. What is the great circle distance that Magellan traveled on his voyage from South America to Asia?**
Write the answer here. __,__ __ __.__ __ __nm

5. Click the **Measure** tool and choose **Clear Measure Totals**.

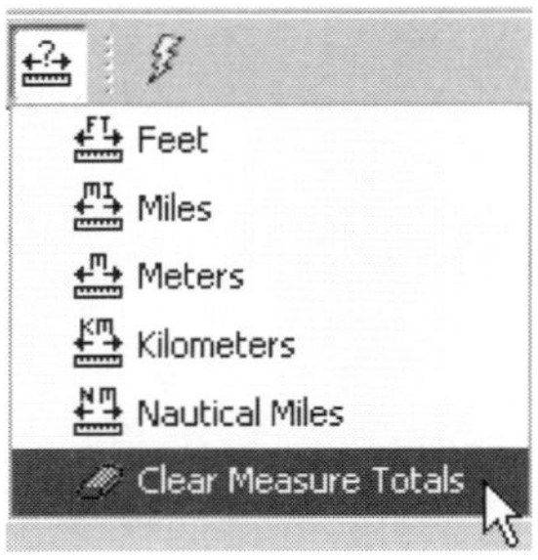

Step 6: Magellan decided to sail to the Philippines instead of the Spice Islands.

At this point, Magellan decided to sail to the Philippines. To follow his route to the Philippines, you will turn the globe again.

1. Click the **Tools** menu and choose **Projection** from the list.
2. Click the **Custom** tab.
3. Click in the **Center Longitude** box, and change the number to **150.0** (**no** minus sign this time).

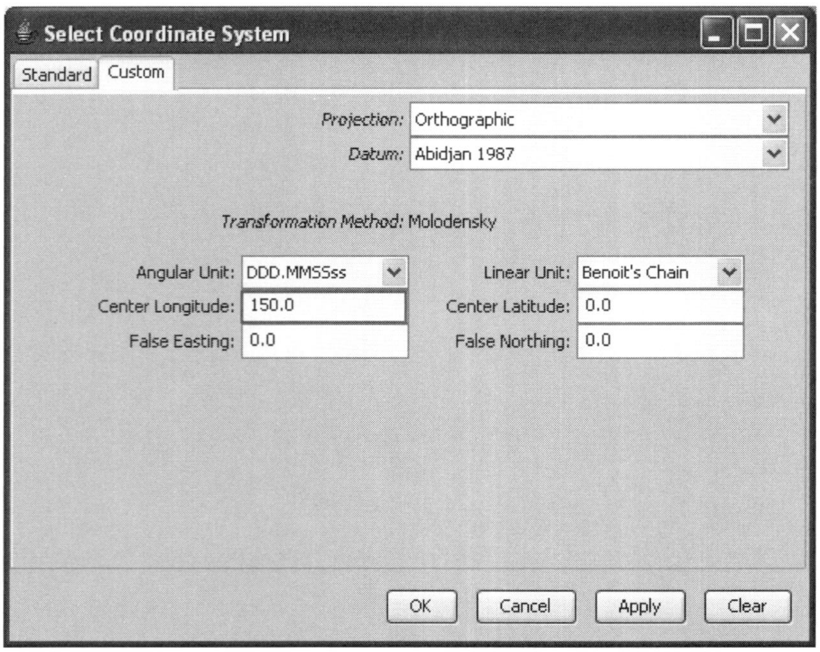

4. Click **OK**.

You will zoom in for a closer look.

5. Turn on the **StudyArea2** layer by clicking the box next to its name.

6. Click the **StudyArea2** layer name to highlight it. It becomes active.

7. Click the **Zoom to Active Layer** button.

8. Turn off the **StudyArea2** layer by clicking the check mark next to its name.

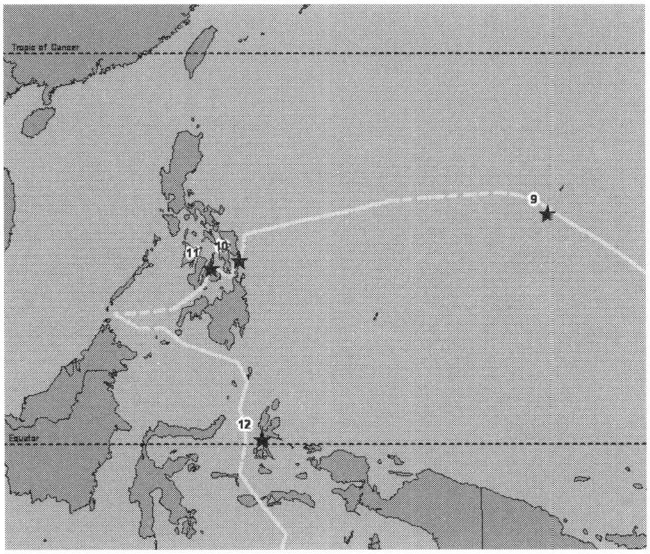

M1 L2

Q8 **How did your map change? (Circle all the correct answers.)**

a. **The map shows more detail (larger scale).**

b. **The map shows less detail (smaller scale).**

c. **The map shows a bigger area.**

d. **The map shows a smaller area.**

On March 16, 1521, four months after leaving the west coast of South America, Magellan reached the Philippine Islands.

Q9 **What is the name of the island at Stop 10? (Hint: Use the MapTip.)**

____

Here the crew gathered and recovered from scurvy when they ate fruit with lots of vitamin C. On March 27, 1521, Magellan realized he was close to the coast of Asia when a large canoe with people brought him gifts made in China.

Step 7: Magellan further explored the Philippine Islands, from Stop 10 to Stop 11.

Q10 **What is the name of the island at Stop 11? (Hint: Use the MapTip.)**

____

In the days that followed, Magellan and his men spent their time trading, relaxing, and converting natives to Christianity. When some rajahs, or chiefs, refused to convert, Magellan decided to punish them but was instead killed during a battle with the native peoples.

His men left Magellan's body behind, and two ships set sail for the Spice Islands. They landed at Stop 12 in November of 1521, after a voyage of 820 days (more than 2 years) since they left Spain. They spent three months in the Spice Islands (also known as the Moluccas), stocking up on spices and preparing to return to Spain.

To see the route back to Spain, you will turn the globe again.

1. Click the **Tools** menu and choose **Projection** from the list.
2. Click the **Custom** tab.
3. Click in the **Center Longitude** box and change the number to **90.0** (**no** minus sign).

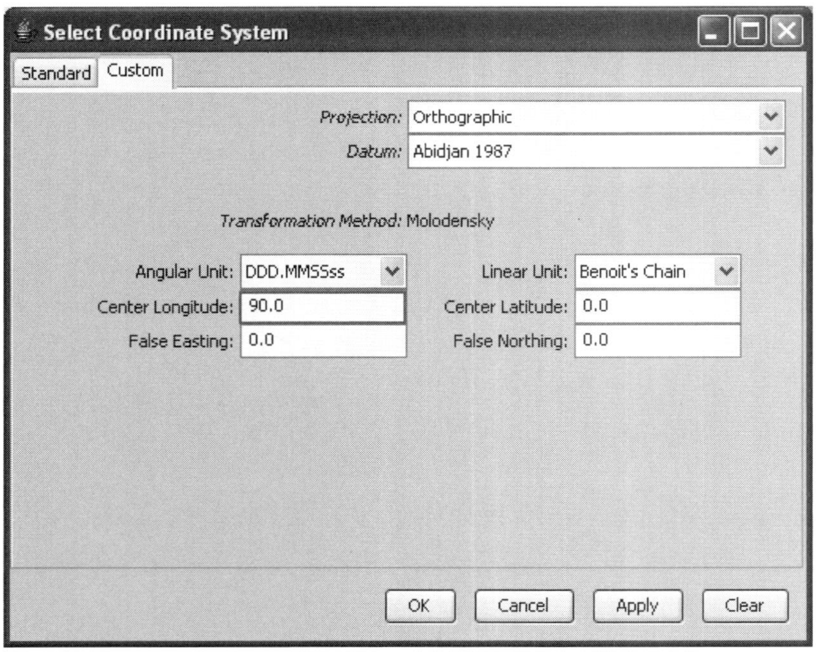

4. Click **OK**.

5. Click the **Zoom to Full Extent** button.

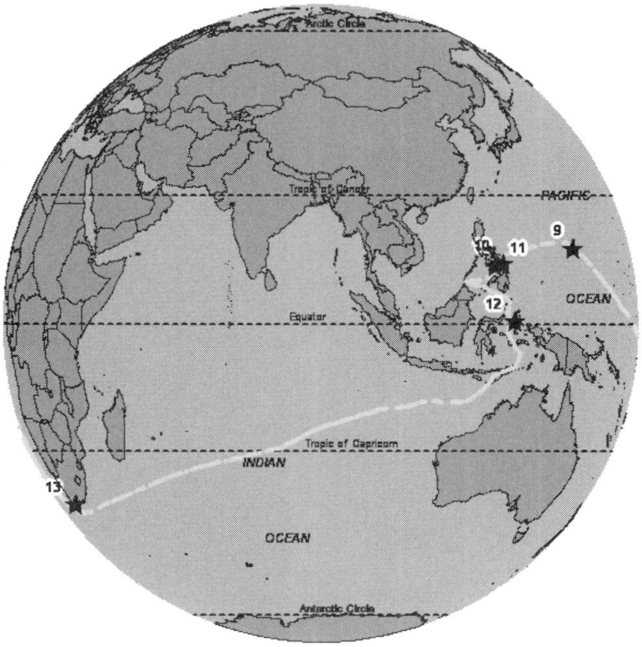

Q11 **What happened to your map? Describe the changes in your own words.**

____

____

____

Q12 **What ocean did they cross on their way back to Spain? (Circle the correct answer.)**

a. Pacific Ocean

b. Indian Ocean

c. Atlantic Ocean

d. Arctic Ocean

6. Turn off **Countries** by clicking the check mark next to its name.
7. Turn on **Continents** by clicking the box next to its name.

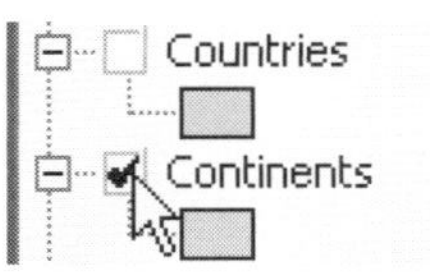

You will find out how far it is from the Spice Islands to the southern tip of Africa.

8. Click the **Measure** tool and choose **Nautical Miles**.

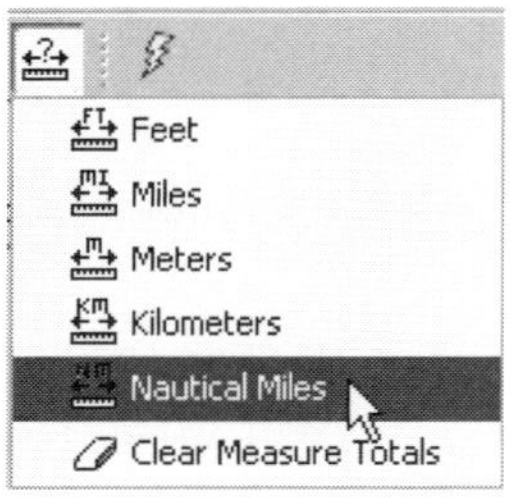

9. Click and *hold* your mouse on Stop 12.
10. Drag your mouse to Stop 13.
11. Let go of your mouse button.

Q13 **Look at the blue geodesic measurement to find out the great circle distance from the Spice Islands to Africa.**
Write it here: __,__ __ __ . __ __ __ nm

Q14 **Compare your measurement in Q7 (across the Pacific Ocean) with the measurement you just made (across the Indian Ocean). Which ocean is the greatest distance across? (Circle the correct answer.)**

a. Indian Ocean

b. Pacific Ocean

12. Click the **Measure** tool and choose **Clear Measure Totals**.

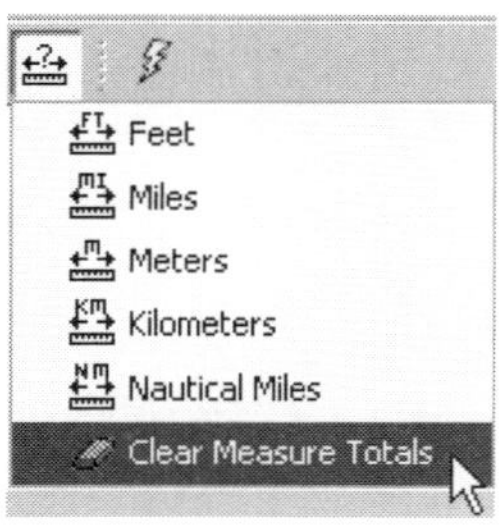

Step 8: Heading for home.

In February 1522, one of the two remaining ships began to sink. Half the men were sent onward under the command of Juan Sebastian del Cano. The others stayed behind in the Spice Islands to make repairs but were soon captured by pirates and hanged.

Del Cano led the crew into the Indian Ocean at the end of March 1522. Soon they were low on water and food and sick with scurvy again. Violent storms damaged the ship. Late in May of 1522, they managed to sail past the cape at the southern tip of Africa (Stop 13).

Q15 What is the name of this cape? (Hint: Use the MapTip.)____________________

By July, they were completely out of food and water. Only 24 men were left, so they headed for the Cape Verde Islands.

M1 L2

To see the route, you will turn the globe one last time.

1. Click the **Tools** menu and choose **Projection** from the list.
2. Click the **Custom** tab.
3. Click in the **Center Longitude** box and change the number to **0.0** (**no** minus sign).

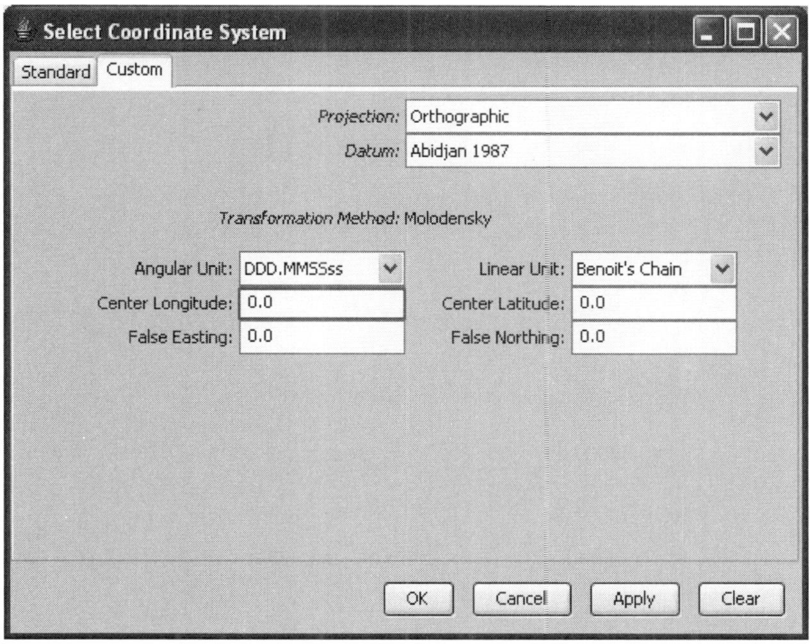

4. Click **OK**.
5. Click the **Countries** layer name to highlight it. It becomes active even though it is not turned on.

6. Click the **Zoom to Active Layer** button.
7. Turn on the **Countries** layer by clicking the box next to its name.

8. Turn off the **Continents** layer.

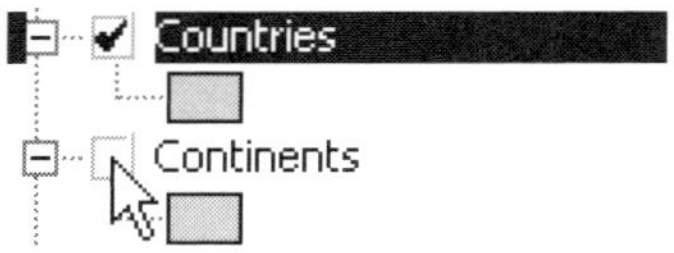

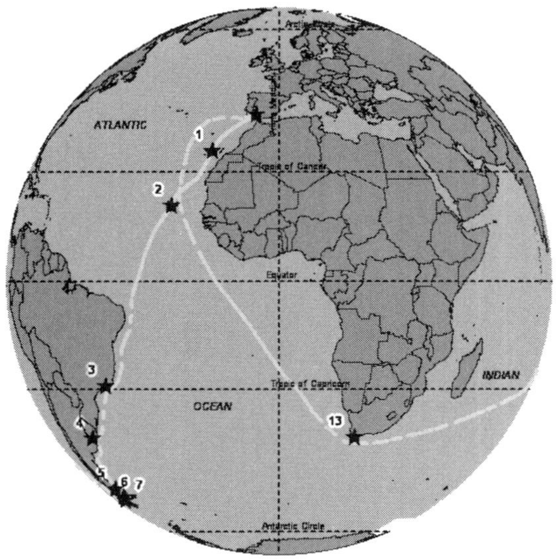

Do you remember where the Cape Verde Islands are? Magellan stopped there on his way to South America.

Q16 **Which stop represents the Cape Verde Islands? (Hint: Use the MapTip.)**

a. Stop 13

b. Stop 3

c. Stop 2

d. Stop 1

Q17 **Which major lines of latitude did they cross on their way to this stop from the tip of Africa? (Circle all the correct answers.)**

a. Antarctic Circle

b. Tropic of Capricorn

c. Equator

d. Tropic of Cancer

e. Arctic Circle

The Cape Verde Islands were in the half of the world claimed by the Portuguese, so del Cano and his Spanish crew were careful. When Portuguese troops arrived from a nearby fort, del Cano left in a hurry, leaving some of his men behind. He was afraid of losing the 26 tons of valuable spices (cloves and cinnamon) his ship was carrying.

In the first week of September 1522, almost three years after the voyage began, del Cano entered the bay that leads to the city of Seville, Spain—where the voyage started. Only 18 men had survived the journey. They were the first men known to travel all the way around the world. We call this a *circumnavigation*.

Q18 **Which major line of latitude did they cross on their way from Stop 2 to their final destination? (Circle the correct answer.)**

a. **Antarctic Circle**

b. **Tropic of Capricorn**

c. **Equator**

d. **Tropic of Cancer**

e. **Arctic Circle**

Q19 **Write a short paragraph explaining what Magellan's expedition accomplished after leaving the South American continent. (Hint: What bodies of water and what major lines of latitude did the expedition cross? What continents and islands did the expedition visit?)**

____

____

____

____

____

Step 9: Save your work and exit AEJEE.

1. Ask your teacher where and how to save your work.
2. Click the **File** menu and choose **Exit**.

Conclusion

Magellan's mission was to find a new route to the Spice Islands and claim them for Spain. Now you know that Magellan was killed and never reached the Spice Islands. Some of his crew made it there and returned with tons of spices, valuable enough to pay for the trip.

After Magellan's fleet visited the Spice Islands in 1521, Spain claimed that these islands were within their (western) half of the world. Eight years later, Portugal paid the Spanish 350,000 gold coins to give up their claim to the Spice Islands.

While few of Magellan's men survived, some of their most important accomplishments include the following:

- Being first to circumnavigate the globe
- Being first to navigate the strait in South America connecting the Atlantic and Pacific oceans
- Observing animals that were entirely new to Europeans (like a llama and a penguin)
- Discovering two galaxies (now called the Magellanic Clouds) in the night sky over the southern hemisphere
- Sailing 14,460 leagues (69,800 kilometers or 43,400 miles) around the earth, discovering its full extent (the actual extent or circumference of the earth is 40,075.16 kilometers or 24,901.55 miles)

Name: ______________________________

Date: ______________________________

Module 2, Lesson 1

Mapping a zoo

Animals are a big part of our life. Animals fascinate us, whether they live with us as pets or roam wild places on our planet. One exciting way to connect with animals from beyond our back yard is to visit the zoo! Zoological parks, or zoos, are a great way to bring people closer to animals. It is a chance for people to more deeply appreciate and understand how animals live and what they are like.

Zoos have been around for a long time. Queen Hatshepsut of Egypt had one about 3,500 years ago, and the Chinese emperor Wen Wang created a large zoo named the Garden of Intelligence about 3,000 years ago. Many leaders used zoos to show power and wealth. Zoos became popular starting about 500 years ago in the 1500s, when European explorers brought animals from the New World (the Americas) back to Europe.

Nowadays, zoos are a place where people can see animals from all over the world. Zoos are also a place where researchers can study animals more closely so that wildlife can survive where they normally live, in their natural habitat. This is important when evaluating how to help certain animals, especially endangered species, that otherwise might not survive on their own.

For this exercise, imagine that a new zoo is being built near your home. The animal enclosures are ready, and the animals are about to move into their new homes. With opening day quickly approaching, the zoo managers realized that they forgot to prepare a map for visitors! Your job will be to create a map that identifies the kinds of animals in the zoo, shows where they live, and provides information about the visitors' center, restrooms, and other facilities.

Step 1: Start AEJEE.

1. Ask your teacher how to start the AEJEE software.
2. Click the **Maximize** button at the top right of the window. Now the AEJEE window fills your screen.

Step 2: Open the project.

1. Click the **Open** button.
2. Navigate to the **OurWorld1\Module2** folder.
3. Choose **Mappingazoo.axl**.
4. Click **Open**.

A map of a zoo will be on your screen with other background information. On the left side of the map, you see a list of layers. Layers are used to show geographic data on a GIS map. Each layer has a name and a legend. You can turn each layer on and off. At the top of the map, you see buttons and tools that you will use during this activity.

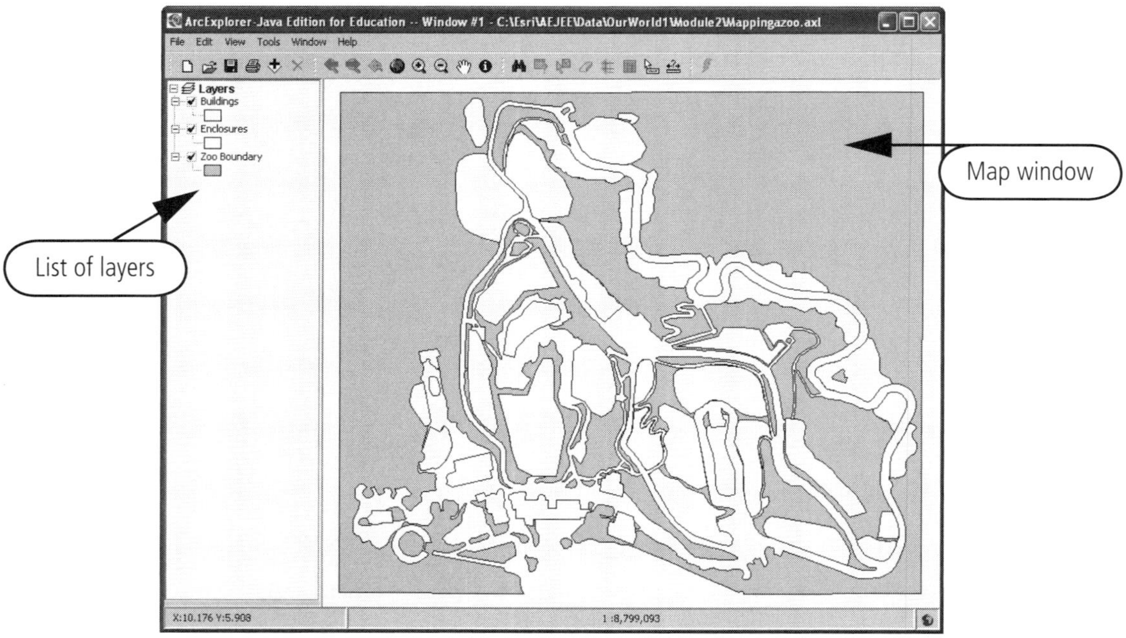

Step 3: Finding things at the zoo.

First you should look at the information that is already on your map. Besides the animals themselves, there are many other features that make up a zoo.

Q1 **Look on your map. Answer the following questions:**

1. **Can you find the restrooms?** **Yes** **No**
2. **Can you tell where the zebra lives?** **Yes** **No**
3. **Can you tell where the tropical rainforest is?** **Yes** **No**

Q2 **All your answers should be No. That's okay. Can you give two reasons why it is hard to find these things?**

1. ______________________________

2. ______________________________

M2 L1

Step 4: Finding the gift shop.

To help visitors find what they are looking for on the map, you should consider the kind of geographic data you can use to improve or enhance your map. Each layer has an attribute table full of information about the features in that layer.

Let's take a closer look at the attribute table for the Buildings layer.

1. Click the **Buildings** layer name to make it active. It will become highlighted.

Note: Making a layer active tells the GIS that this is the layer you want to work with or learn about.

2. Right-click the **Buildings** layer and choose **Attribute Table**. The attribute table appears.

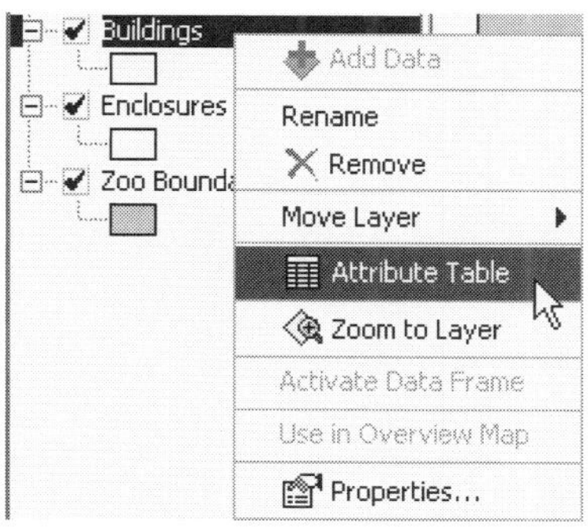

An attribute table is made up of fields (columns) and records (rows). This is where all the information about a layer is contained.

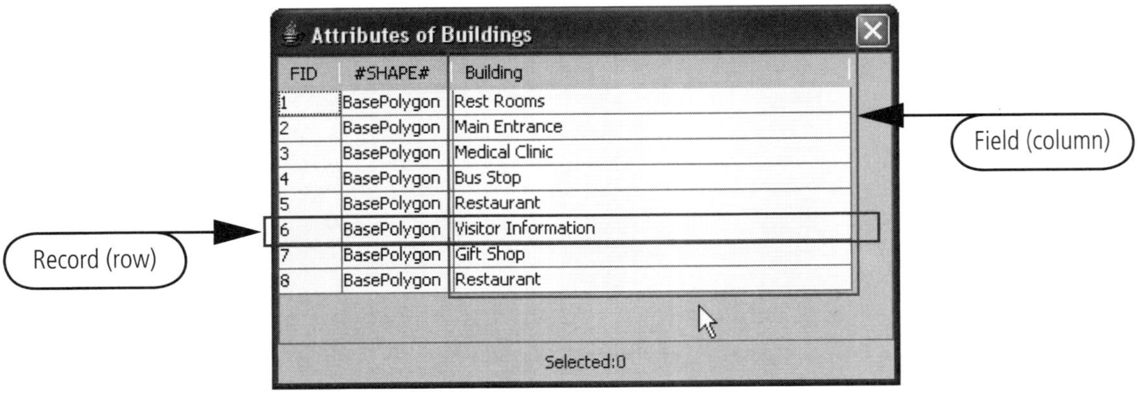

Q3 **How many columns (fields) are in the table? Count them on your screen. You do not need to click the table.**

a. 5

b. 3

c. 2

Q4 **How many rows (records) are in the table? Count them on your screen. You do not need to click the table.**

a. 8

b. 6

c. 9

Q5 **Look under the Building field. Write down the name of two buildings that are listed in the attribute table.**

1. ______________________________

2. ______________________________

3. Close the **Attributes of Buildings** window.

Now you can use the information you found in the attribute table to add more details to your map.

4. Right-click the **Buildings** layer and choose **Properties**.

5. On the **Symbols** tab, change the Color to **Gray**.

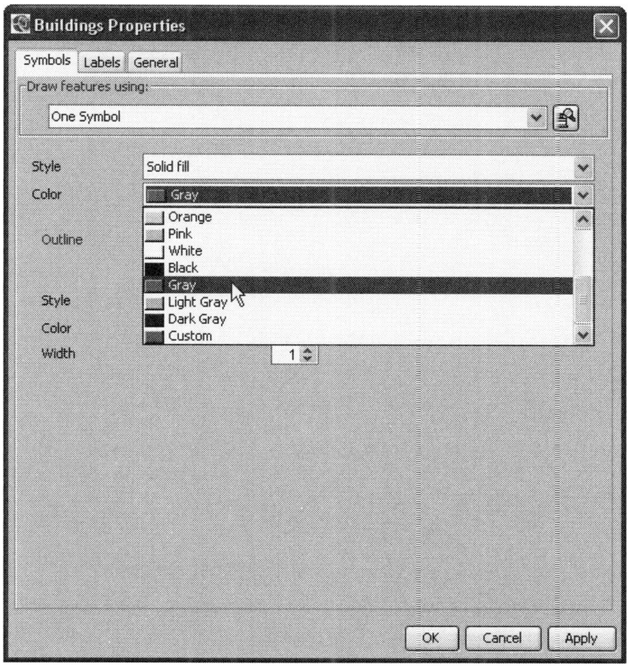

6. Click **Apply**.
7. Move the **Building Properties** window out of the way to see more of the map.

Now the buildings are all colored gray. But you still don't know which building is which. You need to add some labels.

8. In the **Building Properties** window, click the **Labels** tab.

M2
L1

9. Under **Label features using**, choose **Building**.

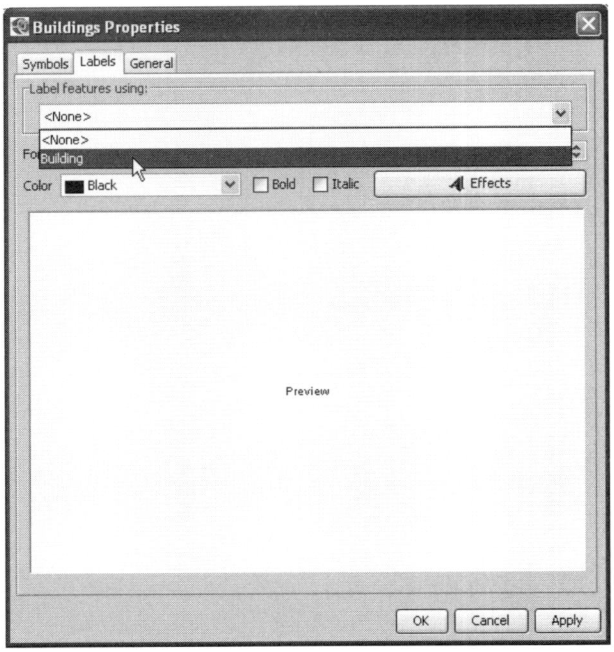

10. Click **Effects**.
11. Put a check mark next to **Glow**.
12. Change the glow color to **Cyan**.

13. Click **OK**.
14. Click **OK** again.

The buildings are now labeled with their name. Visitors can find their way to the gift shop, the restrooms, and the restaurants.

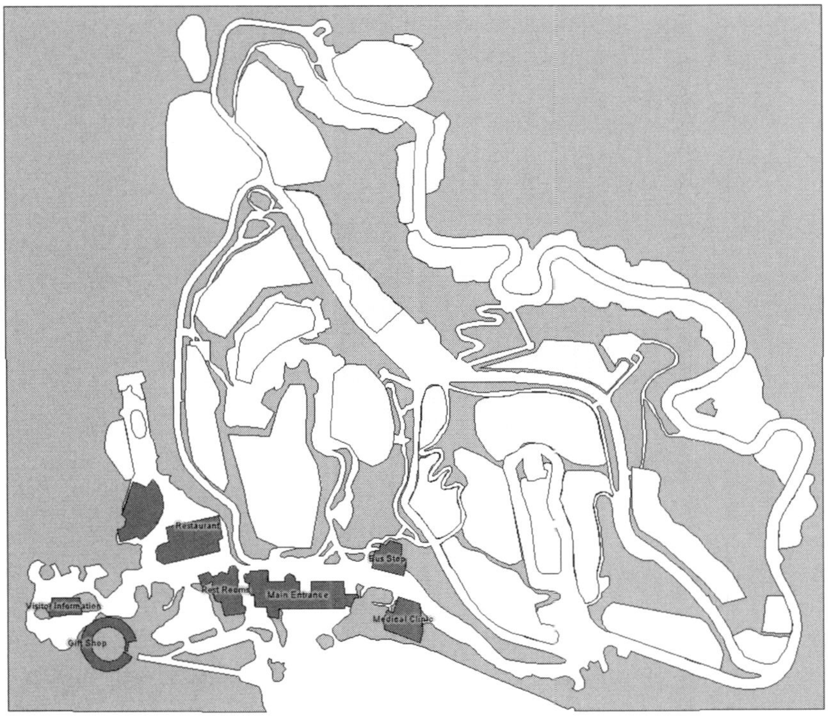

15. Ask your teacher about where and how to save your work. Save your project as **Mappingazoo1.axl**.

Step 5: Finding the tropical rainforest.

Zoos increasingly try to build their enclosures to imitate the natural habitat of animals that live in the wild.

We are going to make a map so that visitors can easily find the tropical rainforest that the zoo created and visit the animals living there. They will also be able to visit all the other types of habitats the zoo created along the way.

1. Click the **Enclosures** layer to make it active. It becomes highlighted.
2. Right-click the **Enclosures** layer and choose **Attribute Table**.

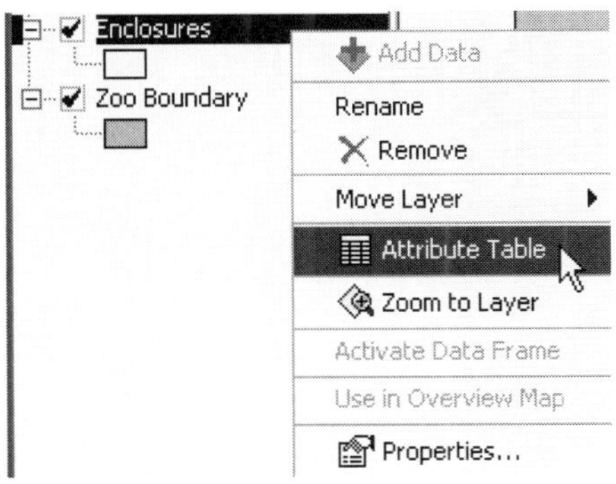

This attribute table contains all the information you have about the enclosures at the zoo. You can organize your table in different ways so that it is easier to read.

3. Right-click the field name **Habitat** and choose **Sort Ascending**. This will rearrange the habitats in alphabetical order.

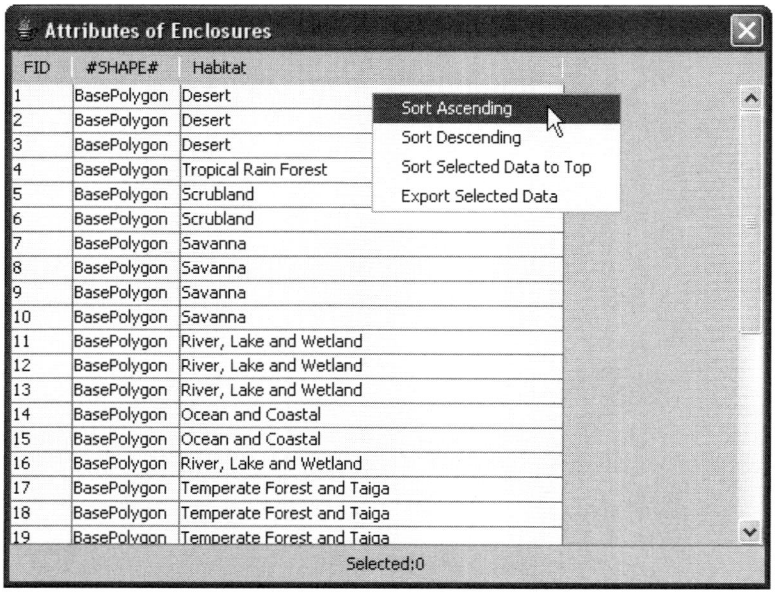

Q6 **What kind of attribute information is available in the Habitat field?**

____

____

Q7 **How many different types of habitat are there?**

a. 8

b. 6

c. 10

Q8 **List two habitats here:**

1. ____________________________________

2. ____________________________________

4. Close the **Attributes of Enclosures** table.

Q9 **Look at your map. Can you see which enclosure represents which habitat on your map? Yes or No**

5. Once again, your answer is No. If the habitats were different colors, they would be a lot easier to see. Let's set up symbols for the enclosures.
6. Right-click the **Enclosures** layer and choose **Properties**.

7. Under **Draw features using**, choose **Unique Symbols**.
8. Next to **Field for values**, choose **Habitat**.

   This will assign a different color to each enclosure based on the habitat type it represents.

9. Next to **Color Scheme**, choose **Minerals**.
10. Keep the **Style** as **Solid Fill**.

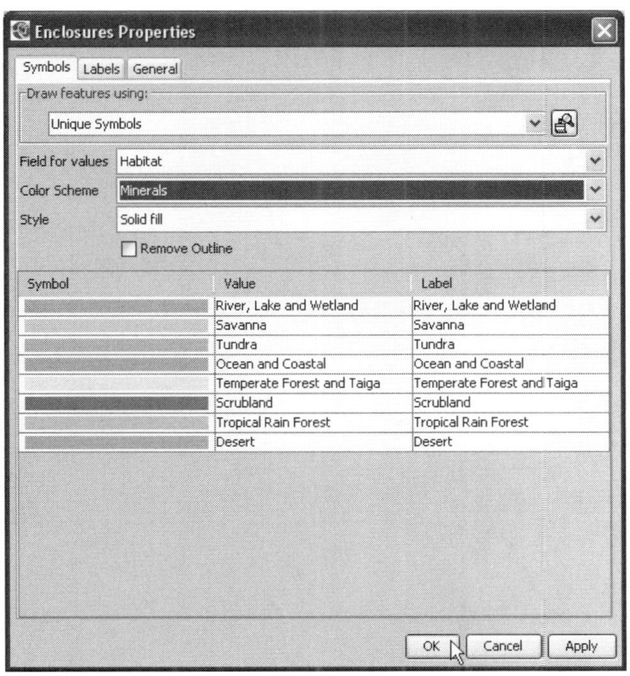

11. Click **OK**.

Q10 **What happened to your map? Describe what happened in your own words.**

____

____

12. In the list of layers, look at the legend for the **Enclosures**.

Q11 **What color is the tropical rainforest on your map?**

____

13. In the legend for **Enclosures**, click the symbol that represents **Tundra**.

All the tundra will be highlighted in green.

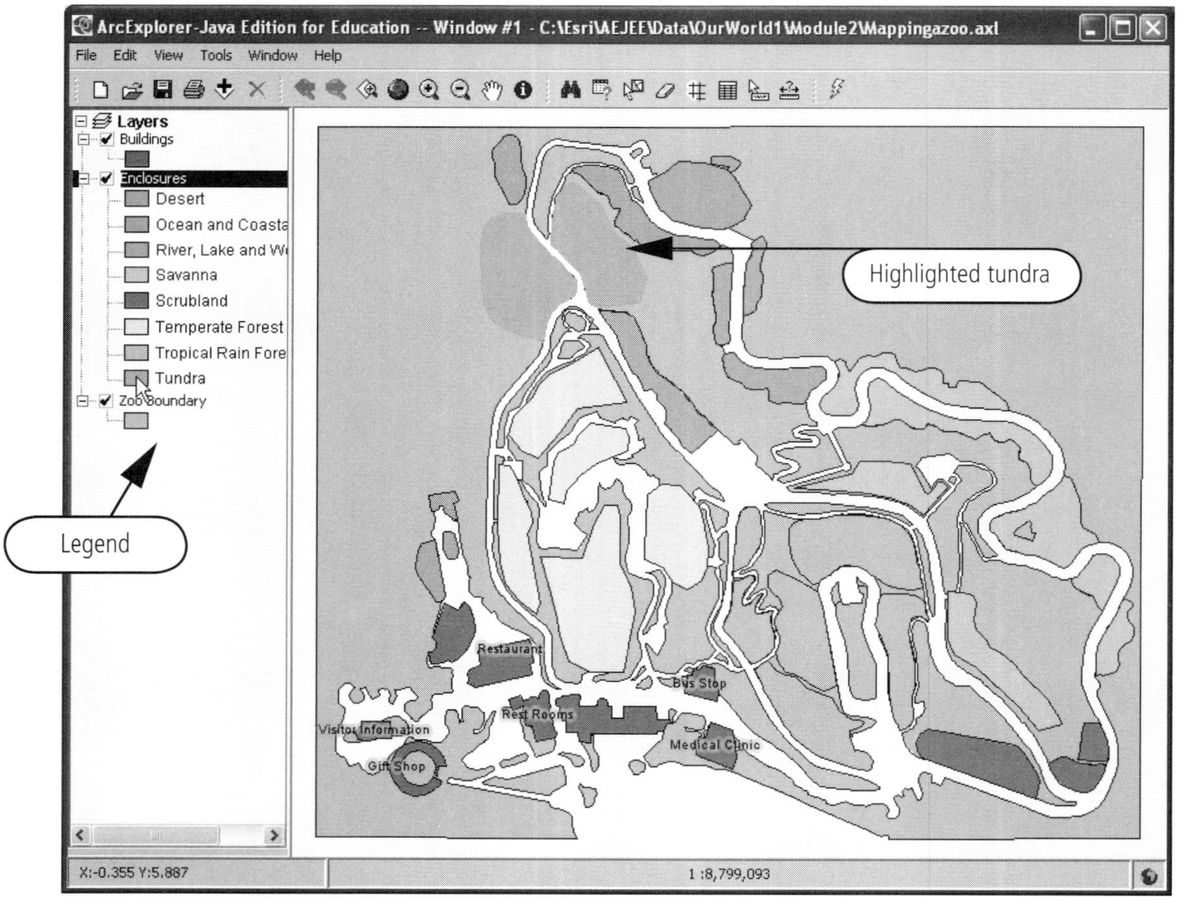

14. In the legend again, click the symbol for **Scrubland**.

All the scrubland areas are now highlighted in green.

Now you can see the relationship between the attribute table and the map. Using the attribute table, let's make your map more useful.

15. Click the **Clear All Selections** button.
16. Ask your teacher about where and how to save your work. Save your project as **Mappingazoo1.axl**.

Step 6: Adding the animals.

Sometimes, a zoo groups animals together based on their appearance. Other times, they are grouped based on their collective activities. Butterflies need to be in enclosed spaces (like a butterfly emporium) so they can't fly away. Monkeys need lots of trees and branches for swinging and climbing. Polar bears need a pool for swimming.

Your map is already full of great information, but you are still missing one key piece—the animals! Let's add them to the map.

1. Click the **Add Data** button.
2. Navigate to the **OurWorld1\Module2\Data** folder.
3. Choose **ZooAnimals.shp**.

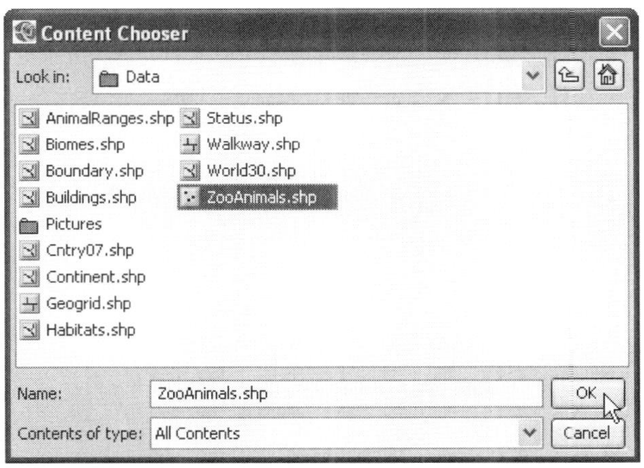

4. Click **OK**.

A new layer appears in the list on the left. A set of points appears on your map. The points represent different species of animals living in the zoo. Each animal is located in its proper habitat. Let's take a closer look at which animals live at the zoo.

5. Click the **ZooAnimals** layer name to make it active.

6. Click the **Identify** tool.

7. Click any animal point. This will bring up the **Identify Results** window.

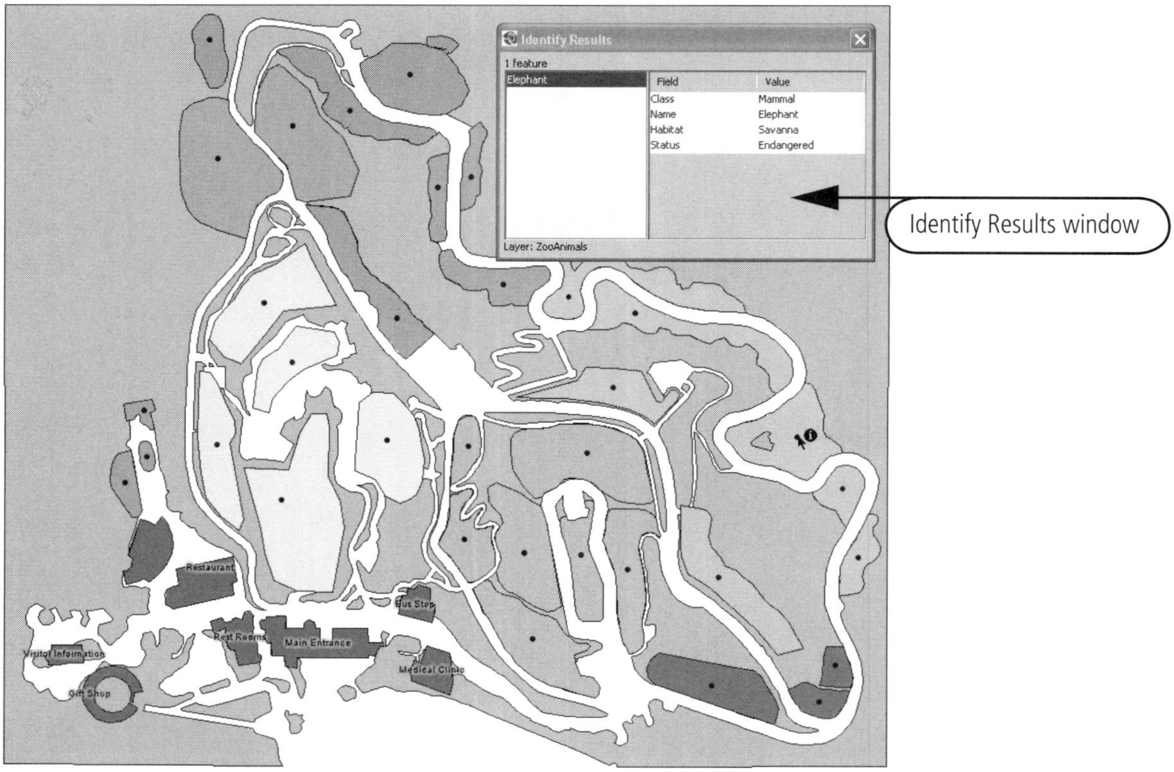

(Q12) Look at the data in the window and complete the following table:

Class	
Name	
Habitat	
Status	

8. Close the **Identify Results** window.

   Using the Identify tool gives you the information for one animal at a time.

9. Right-click the **ZooAnimals** layer and choose **Attribute Table**.

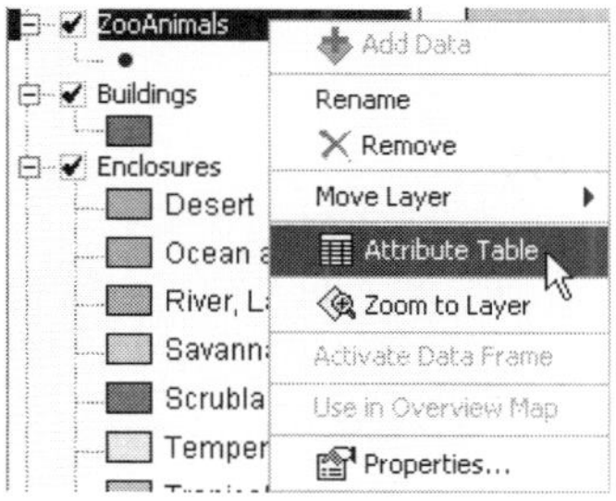

This is the attribute table associated with the animals. It gives you all the information about all the features at once.

10. Right-click the **Name** field and choose **Sort Ascending**. This will rearrange the animals in alphabetical order.
11. Scroll down the list of animals.

M2
L1

Q13 **Based on the attribute data in the table, list one way that you could classify (put into categories) the animals in the zoo.**

____

Q14 **Based on the attribute data in the table, list one way that you could symbolize (represent with colors or symbols) the animals in the zoo.**

____

Q15 **Can you think of another way that animals can be described or classified that is not in this table?**

____

12. Close the **Attribute Table**.
13. Right-click the **ZooAnimals** layer and choose **Properties**.

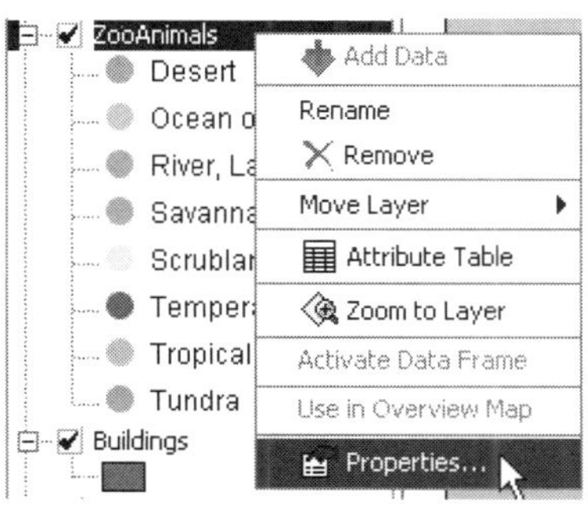

14. Under **Draw features using**, choose **Unique Symbols**.
15. Next to **Field for values**, choose **Habitat**.
16. Next to **Color Scheme**, choose **Minerals**.
17. Leave the **Style** as **Circle**.
18. Change the **Size** to **12**.
19. Click **OK**.

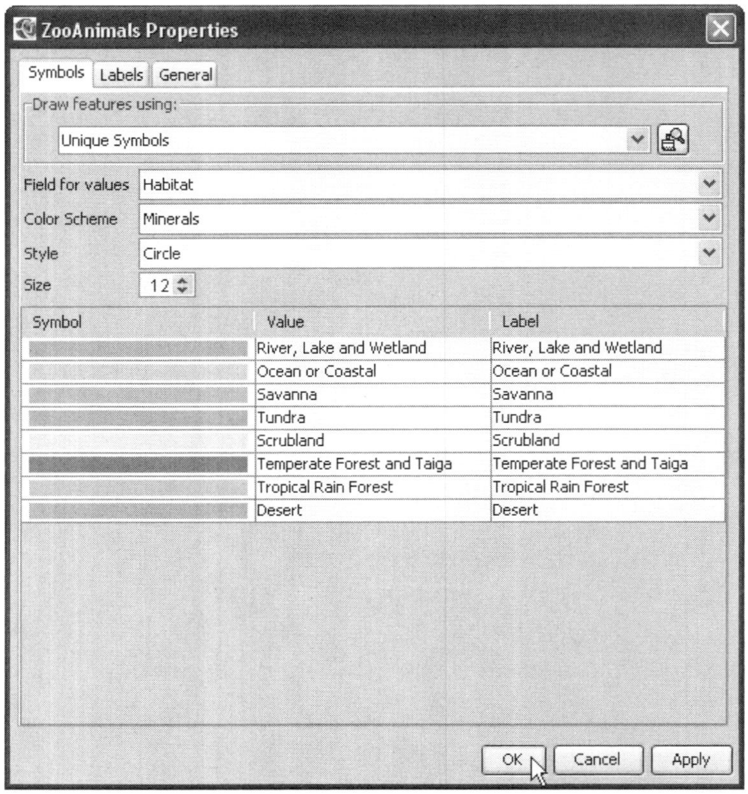

Now the animals are classified using habitat type. This is the same way the enclosures are classified. Does this give a visitor much extra information? Not really.

For example, look at the savanna habitat area. What information are you showing about the animals that live in that area? Only that they live in the savanna.

Let's try mapping the animals using different information. You can use the Class field so that visitors can see which animals are mammals, reptiles, birds, or insects.

20. Right-click the **ZooAnimals** layer and choose **Properties**.

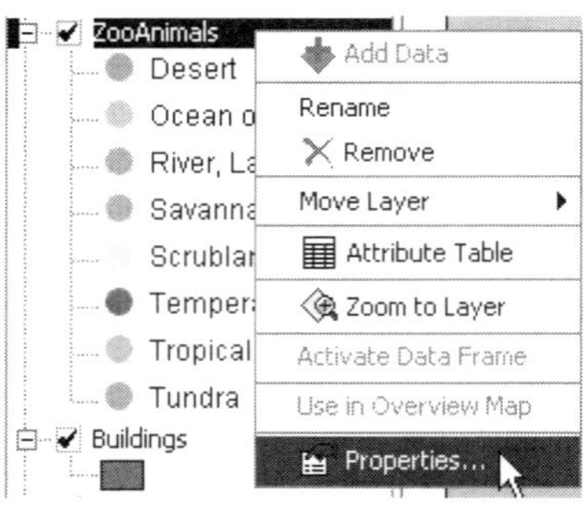

21. Under **Draw features using**, choose **Unique Symbols**.

22. Next to **Field for values**, choose **Class**.
23. Leave the **Color Scheme** as **Random**.
24. Change the **Size** to **12**.

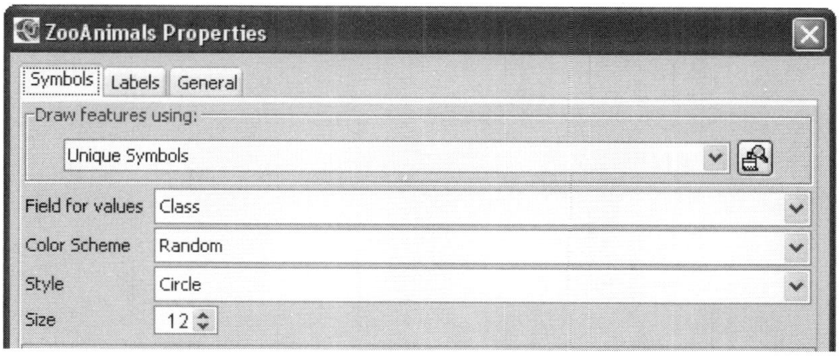

Now you will see four classes of animals with different colors assigned to them. You want to make sure that your colors stand out, so let's make some changes.

25. With the **Properties** window still open, click the color box next to **Bird**.
26. In the color table, choose a **dark blue** color.
27. Click **OK**.

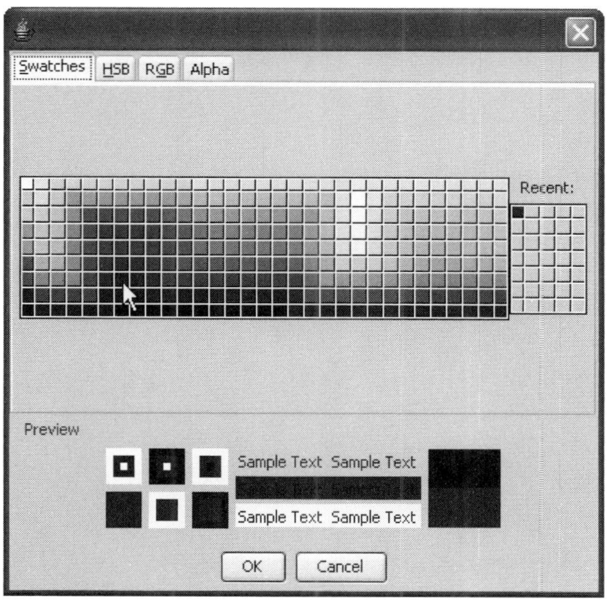

28. Click the color box next to **Insects and Spiders**.
29. In the color table choose a **red** color.
30. Click **OK**.
31. Click the color box next to **Mammal**.

M2
L1

32. In the color table, choose a **green** color.
33. Click **OK**.
34. Click the color box next to **Reptiles**.
35. In the color table, choose a **orange** color.
36. Click **OK**.

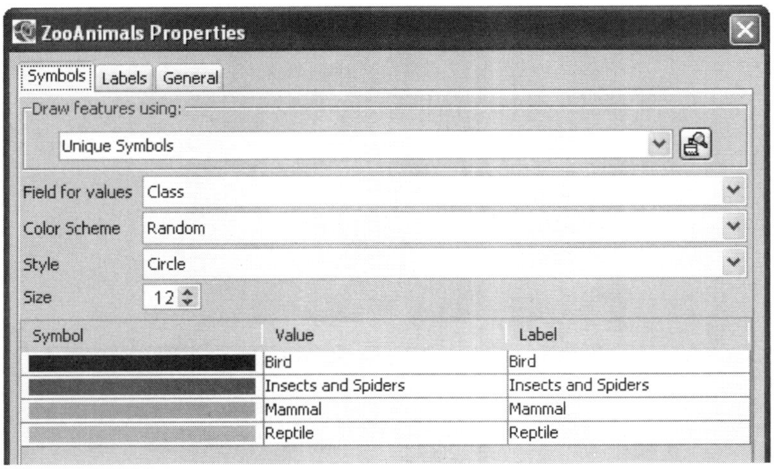

Finally, you need to label the animals so that visitors can see and find the location of specific animals in the zoo.

37. Click the **Labels** tab.
38. Under **Label features using**, choose **Name**.
39. Change the **Font** to **Verdana**.
40. Leave the size.
41. Put a check mark next to **Bold**.
42. In the **Placement** box, click the dot at the top of the circle.
43. Click **OK**.

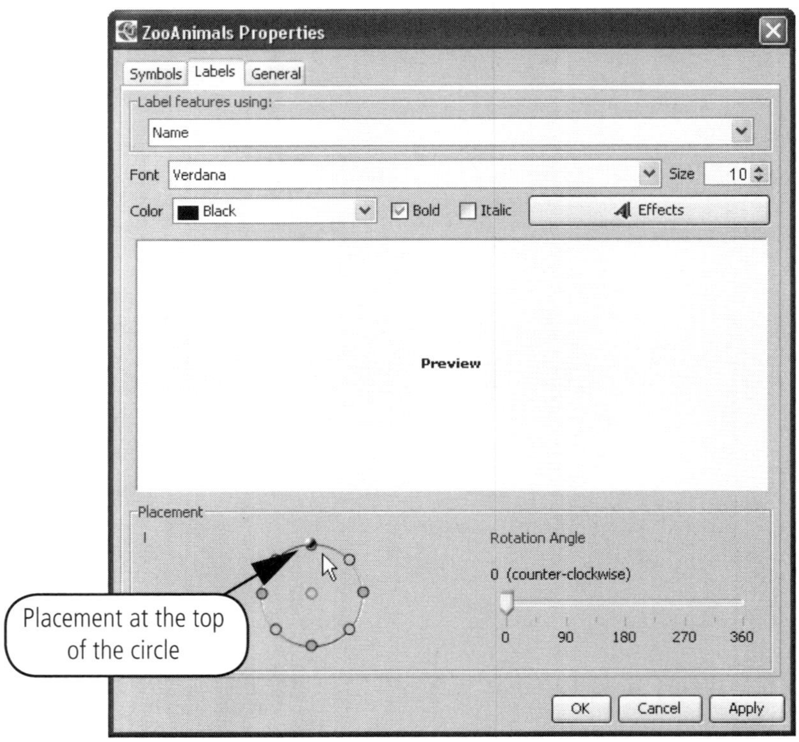

Now the animals are all labeled based on their names.

44. Ask your teacher about where and how to save your work. Save your project as **Mappingazoo1.axl**.

> **Note:** You may notice that some animals are missing a label. This can happen when two labels overlap. You can use the **Identify** tool to find the names of animals without labels.

Q16 **Explain why your map is much more useful to visitors. What does your map show? What can you say about the colors and symbols?**

____

____

____

Q17 **Go back to step 3. Can you answer Yes to those questions now?**

a. Yes

b. No

M2
L1

Step 7: You are ready for opening day!

Your new map is now informative, helpful, and also looks really nice. It's ready for visitors to use on opening day. All you need to do is print it.

1. Under the **View** menu, choose **Layout View**. Your map will appear centered on a "paper."

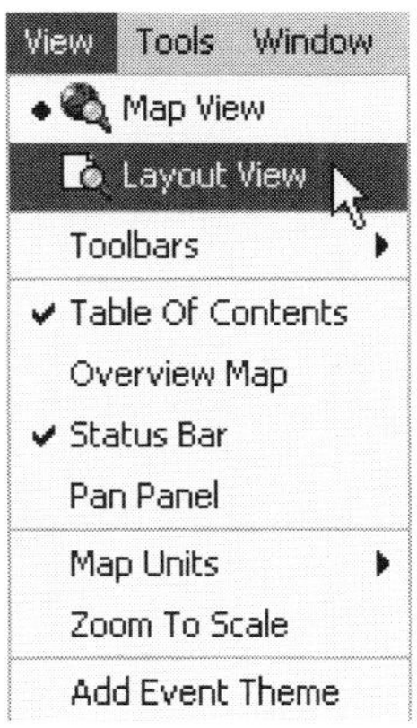

First you need to add a title to your map.

2. Click the **Add text** tool.

You see a white box that says **Right-click this text** in the middle of your map.

3. Click, hold, and drag the box to the top of your layout.
4. Right-click the box and choose **Properties**.
5. Click the white box and write the name of your zoo here (you can choose any name you want!).
6. Click **Change Properties**.
7. Change the **size** to **44**.
8. Click **OK**.
9. Click **OK** again.

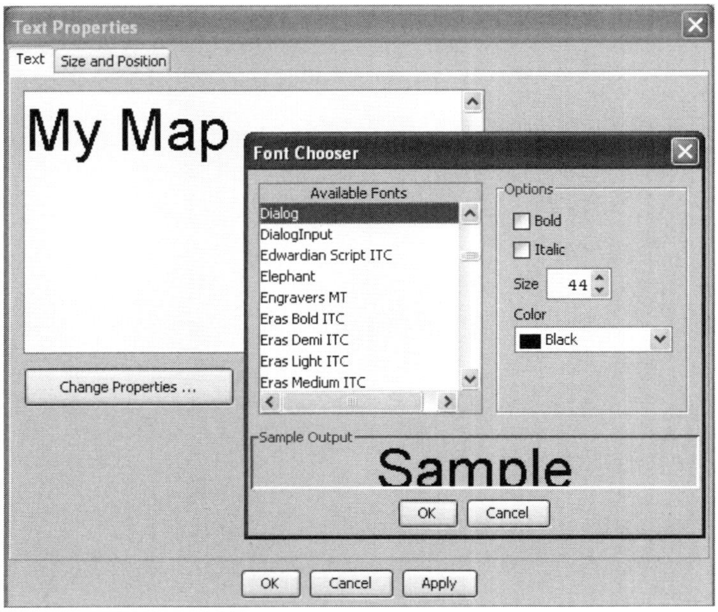

10. Click and hold on your title to move it around. Place it where you want it.

Now you need to add a legend to the map so that people can understand the way you classified things.

11. Click the map frame once. Blue dotted lines appear around it.
12. Click the **Add map legend** tool. Your legend appears on the map.
13. Click, hold, and drag your legend to the bottom right of your layout. You can resize the legend if you want.
14. Click the **Add image** button.
15. Navigate to **OurWorld1\Module2\Data\Pictures**.

16. Select ONE picture from the list of animals:
    - Elephant
    - Flamingo
    - Giraffe
    - Gorilla
    - Hippopotamus
    - Lion
    - Panda
    - Tiger
    - Zebra
17. Click **Open**.

    The animal will appear in a box on your map (it might appear right in the middle of your map).

18. Click, hold, and drag the box to the empty space on the top right corner of your page. You can resize the box if you want.
19. You can rearrange all the different parts, or elements, on your map until you are happy with the way it looks.

M2
L1

Step 8: Save your work, print your map, and exit AEJEE.

1. Ask your teacher about where and how to save your work. Save your project as **Mappingazoo1.axl**.

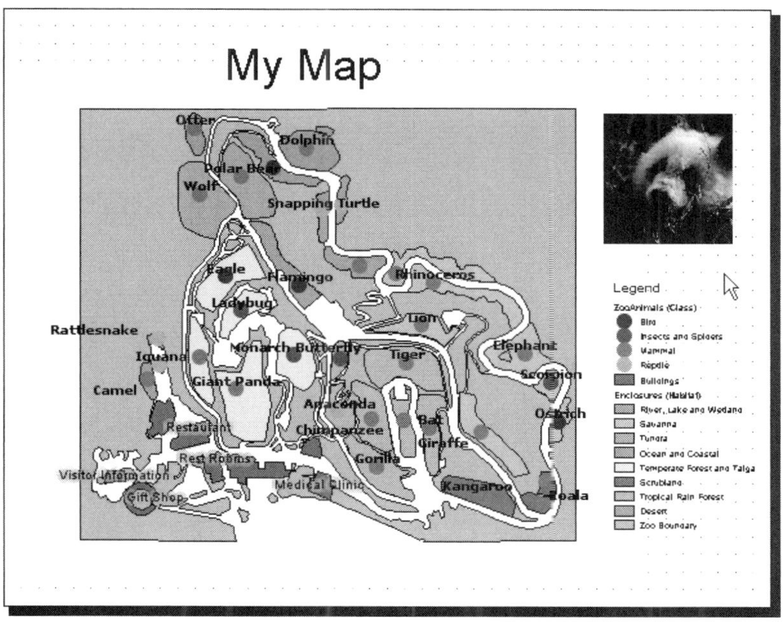

2. Ask your teacher how to print your map.

3. When you are finished printing your map, click the **File** menu and choose **Exit**.

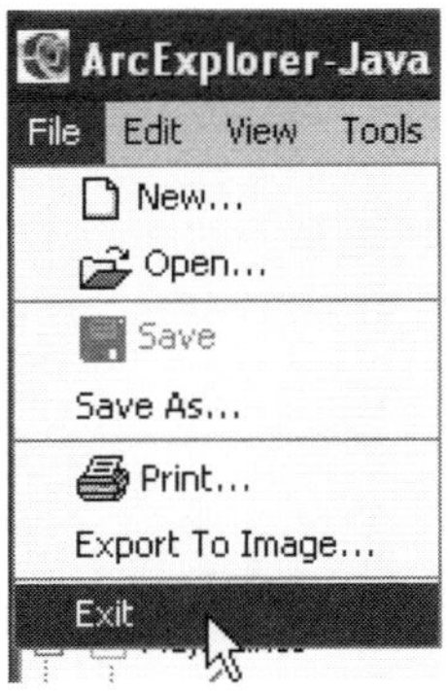

Conclusion

Well done. You have now created a useful and informative map of your community's new zoo. Classifying features is not easy, and it is important to keep in mind what you are trying to show.

Visitors can use your map to plan a full day of fun at the zoo, including observing the animals, having lunch, and shopping. Looking at the map, visitors can learn a lot about the animals living in the zoo and understand how it is organized.

Zoos are constantly evolving and changing. Many animals are losing their natural habitats as the human population increases. Zoos and conservation groups around the world are using geographic information systems (GIS) to locate and monitor habitat loss. GIS is one of many tools that zoos use today to help people understand and appreciate the value of animals on our planet.

Name: ______________________________

Date: ______________________________

Module 2, Lesson 2

Touring a zoo

M2 L2

The animal kingdom is quite large, with thousands of animal species identified around the world and more being discovered all the time. To make sense of all these species, scientists typically classify animals based on their physical characteristics. They start with a general classification and then get more detailed until they end up with a scientific name for the animal. For example, in the Linnaeus classification, the scientific name for a brown bear is *Ursus arctos*. This means that it has a backbone, is a mammal and a carnivore, and is part of the bear family.

Usually, it is easier to use common names to identify animals. In addition to their physical features, animals have many other characteristics: What country or area do they come from? What habitat do they live in? What kind of food do they like to eat?

For this lesson, the new zoo that has opened in your community needs your mapping expertise again. The zookeepers are so pleased with the map you made that they would like you to create a custom visitors tour. Sometimes, visiting a zoo can feel overwhelming because there is so much to see. In the last lesson, you used the animals' class (whether they were mammals, reptiles, or birds) to tell them apart. You will now explore some of the other information you have about the animals.

The zookeepers would like you to make sure visitors see a wide range of animals from different habitats and different classes. Here is a list of the characteristics of the animals that the zoo would like you to include in your tour:

- One reptile
- One bird
- One insect
- One animal that lives in the scrubland
- One mammal that lives in the tundra
- One mammal that lives in the savanna and is endangered

The tour also requires that you stop for lunch at the restaurant and make time to buy some souvenirs in the gift shop.

Let's get started on your tour map!

Step 1: Start AEJEE.

1. Ask your teacher how to start the AEJEE software.
2. Click the **Maximize** button at the top right of the window. Now the AEJEE window fills your screen.

Step 2: Open the project.

1. Click the **Open** button.
2. Navigate to the **OurWorld1\Module2** folder.
3. Choose **TouringtheZoo.axl**.
4. Click **Open**.

   A map of a zoo will be on your screen with other background information. It looks like the map you created in the previous activity. On the left side of the map, you see a list of layers. Layers are used to show geographic data on a GIS map. Each layer has a name and a legend. You can turn each layer on and off. At the top of the map, you see buttons and tools that you will use during this activity.

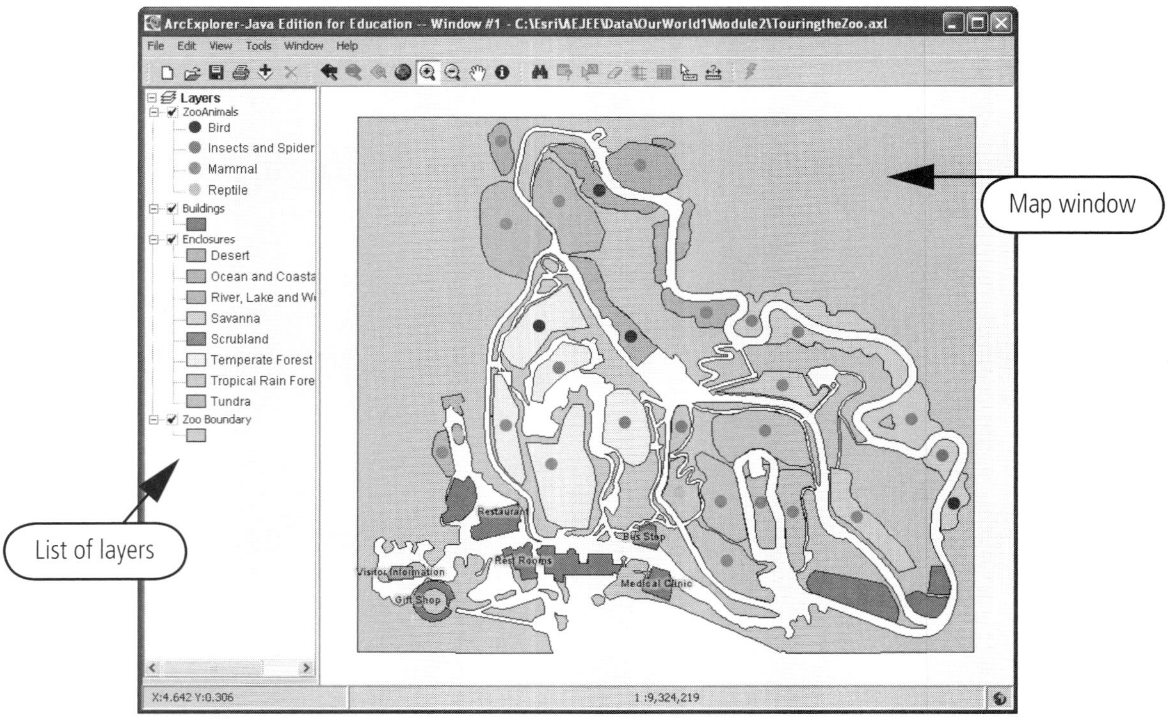

Step 3: Snakes and lizards.

Reptiles are cold-blooded animals with scaly skin. Their skin helps keep their bodies from drying out in the heat. They sit in the sun to get warm and then move to the shade to cool down. This class of animals lays eggs to have their babies. There are more than 6,500 species of reptiles in the world. That's a lot of reptiles!

Remember that your tour needs to include one reptile. Let's see what reptiles are at the zoo so we can choose which one to visit on our tour.

Q1 **Look at the map legend. What is the color for reptiles?** ____________________

A few reptiles live at the zoo. The problem is you don't know what kind of reptiles they are. That information is stored in the attribute table.

1. Click the **ZooAnimals** layer name to make it active.

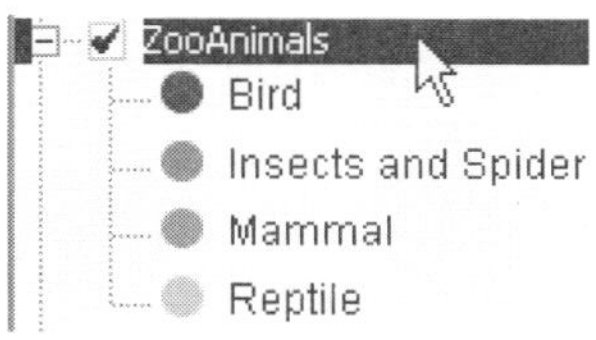

2. Click the **Identify** tool.
3. Click one of the **reptile symbols** in the zoo. This brings up the **Identify Results** window.

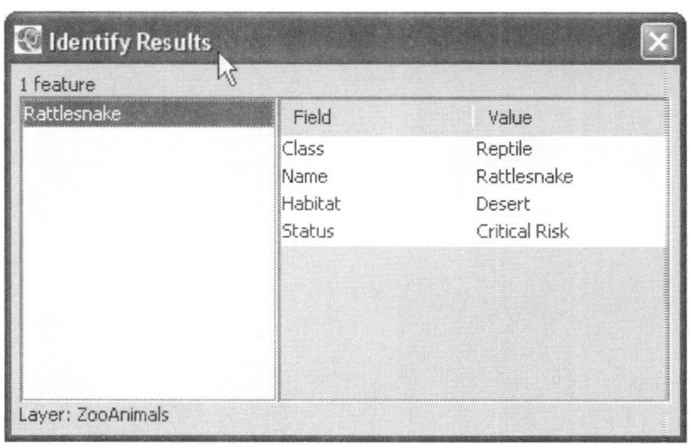

The **Identify Results** window contains information about the feature you selected. The information is coming from the attribute table.

Q2 **Write down the name of the reptile in the Query answers column on your worksheet (worksheet 1).**

4. Close the **Identify Results** window.
5. Click another reptile symbol.

Q3 **Write down the name of this reptile in the Query answers column on your worksheet. Do this for all the reptiles. (Hint: There are 5 of them.)**

6. Close the **Identify Results** window when you are done. You should have a list of all the reptiles in the zoo on your worksheet.

Q4 **Have you ever seen one of these reptiles in pictures or in real life? Write down the ones you have seen.**

____

Q5 **Write down some things you know about one of the reptiles.**

____

____

Step 4: Birds of a feather.

The zoo now wants visitors to see a bird. Birds are warm-blooded animals like mammals, but birds are covered in feathers. They have a backbone and a skeleton. Did you know that some of their bones are hollow to make it easier to fly? Birds lay eggs in nests, and when the babies hatch, the parents bring them food and care for them until they are old enough to fend for themselves. Of more than 8,800 known species of birds, the smallest is the hummingbird, and the largest is the ostrich.

You need to choose one bird for your tour. Do you know what kinds of birds are in the zoo? Let's find out.

Q6 **Look at the legend. What is the color for birds?** ____________________

Instead of using the Identify tool, you are going to select all the birds and then look at the attribute table to see what information you can find.

1. Look in the list of layers. Find the legend for ZooAnimals.

2. Click on the **Bird symbol** in the legend. (Hint: Click the circle.)

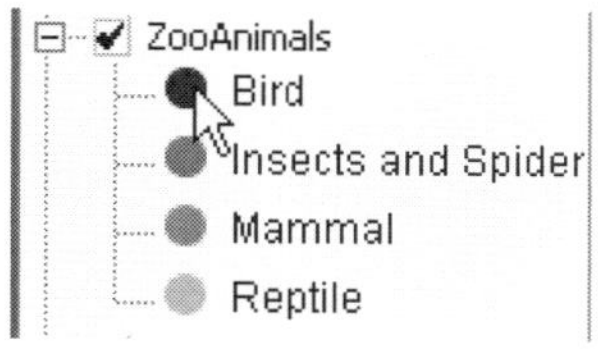

On the map, the bird symbols now have a green dot in the center that indicates they are selected.

3. Look at your map and locate the birds.

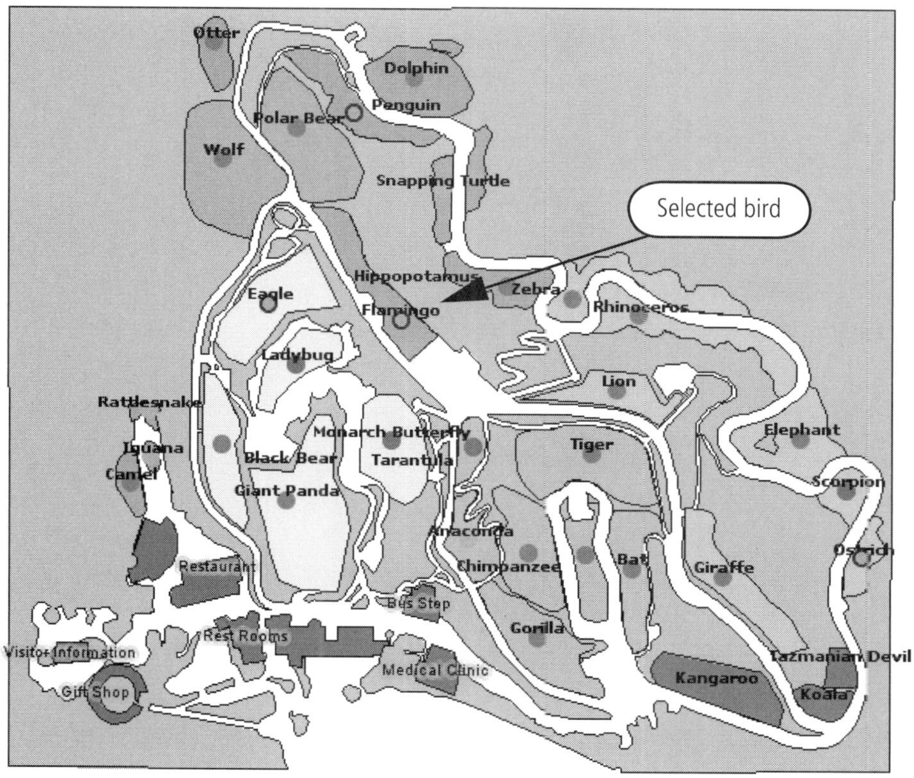

4. Right-click the **ZooAnimals** layer and choose **Attribute Table**. This will open the attribute table for the zoo animals. You will see that the selected features are highlighted in blue.

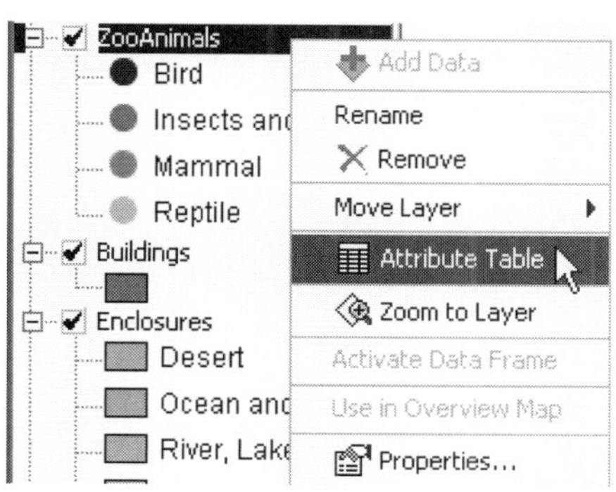

M2
L2

5. Right-click the **Name** field heading and choose **Sort Selected Data to Top**.

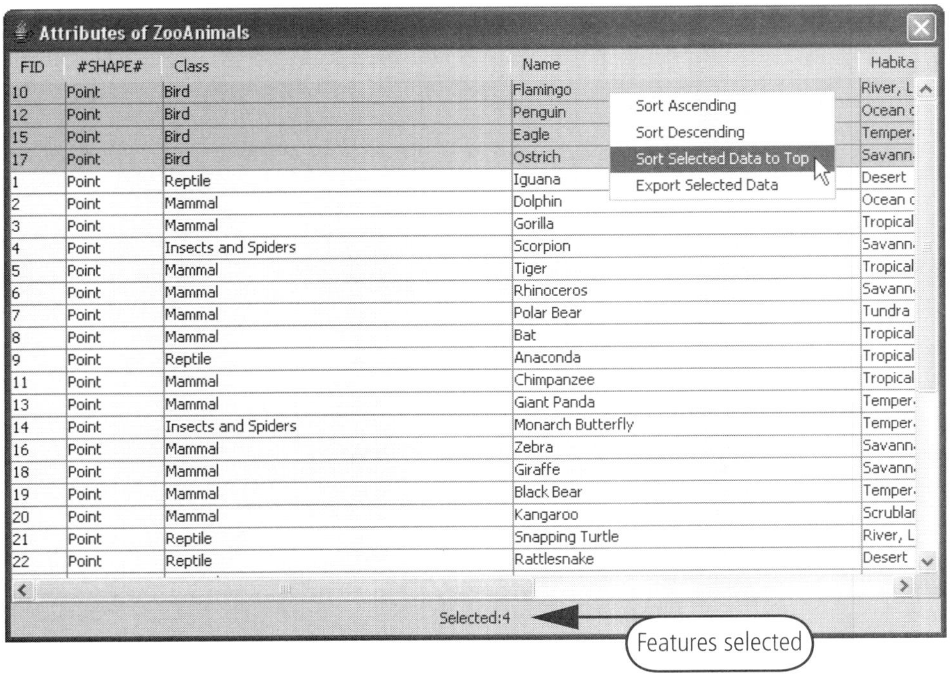

FID	#SHAPE#	Class	Name	Habita
10	Point	Bird	Flamingo	River, L
12	Point	Bird	Penguin	Ocean c
15	Point	Bird	Eagle	Temper.
17	Point	Bird	Ostrich	Savann.
1	Point	Reptile	Iguana	Desert
2	Point	Mammal	Dolphin	Ocean c
3	Point	Mammal	Gorilla	Tropical
4	Point	Insects and Spiders	Scorpion	Savann.
5	Point	Mammal	Tiger	Tropical
6	Point	Mammal	Rhinoceros	Savann.
7	Point	Mammal	Polar Bear	Tundra
8	Point	Mammal	Bat	Tropical
9	Point	Reptile	Anaconda	Tropical
11	Point	Mammal	Chimpanzee	Tropical
13	Point	Mammal	Giant Panda	Temper.
14	Point	Insects and Spiders	Monarch Butterfly	Temper.
16	Point	Mammal	Zebra	Savann.
18	Point	Mammal	Giraffe	Savann.
19	Point	Mammal	Black Bear	Temper.
20	Point	Mammal	Kangaroo	Scrublar
21	Point	Reptile	Snapping Turtle	River, L
22	Point	Reptile	Rattlesnake	Desert

At the bottom of the window, the number of selected features is shown.

Q7 **Write the names of all the birds under the Query answers column on your worksheet.**

Q8 **Scroll your table to the right. Look in the Status field. Which birds are listed as vulnerable (facing a high risk of extinction in the wild)?**

1. ______________________________

2. ______________________________

3. ______________________________

Q9 **Have you ever seen one of these birds in pictures or in real life? Write their names here.** ______________________________

____

Q10 **Write down some things you know about one of the birds.** ______________________________

____

6. Close the **Attribute Table**.

7. Click on **Clear All Selections**.

Step 5: Animals that live in the scrubland.

Scrublands have hot, dry summers and cool, moist winters. The few trees that grow there do not get very big. Although this seems like a harsh habitat, bobcats, leopards, owls, warthogs, and many other kinds of animals live in scrublands. In this zoo, several animals represent the scrubland. You need to choose one of them to add to your tour.

You have already used the Identify tool and selected features using the legend symbol. In a GIS, you can also build queries to get information from the attributes you have.

1. Make sure the **ZooAnimals** layer is still active. (Hint: It should be highlighted in dark blue in the legend.)

2. Click the **Query Builder** button. This opens the Query Builder window.

3. Under **Select a field**, click **Habitat**.

4. Click the = (equals) button.

5. Under **Values**, click **Scrubland**.

   Your query should look like this:
   (**Habitat** = **'Scrubland'**)
   You are asking the GIS to show you all the animals that live in the scrubland.

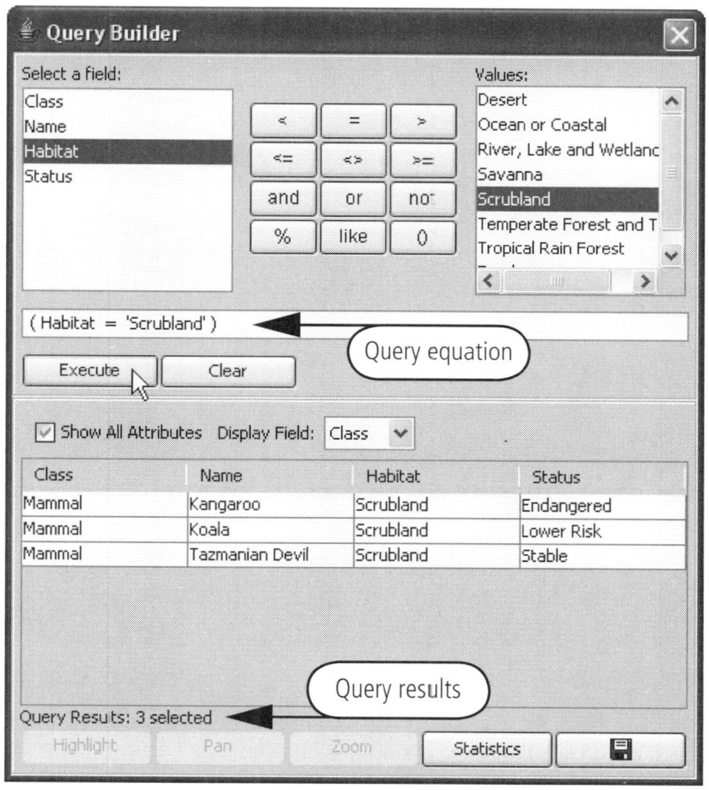

6. Click **Execute**. A list of animals that live in scrubland appears in the lower half of the window.

(Q11) **How many animals were selected?** ______________________________

Q12 **Write the names of all the animals under the Query answer column on your worksheet.**

Q13 **Have you ever seen one of these animals in pictures or in real life? Write their names here.**

Q14 **Write down some things you know about one of the animals.**

7. Close the **Query Builder**.

   You will see the same animals in the list highlighted in green on the map.

8. Click **Clear All Selections**.

Step 6: Mammals and cold climates.

Mammals are animals that grow fur on their bodies and give birth to living young (they do not lay eggs). They also produce milk to feed their babies. Mammals are warm-blooded animals that regulate their body temperatures. However, that doesn't always mean they live in warm places.

Some mammals (and other animals) live in areas called the tundra. Tundra regions are found in the Arctic where temperatures drop as low at -76 degrees Fahrenheit! The summers can be quite warm, but they don't last long. Plants can still grow there, even though a layer of permafrost (ground that stays frozen all the time) covers the ground.

People are fascinated by animals that can survive in this kind of habitat. Let's find out what mammals at the zoo live in tundra regions.

1. Make sure the **ZooAnimals** layer is still active. (Hint: It should be highlighted in dark blue in the list of layers.)

2. Click the **Query Builder** button. This opens the Query Builder window. This time you will build a complex query where you ask the GIS more than one question at a time.

3. Click **Clear** to erase the old query.

4. Under **Select a Field**, click **Class**.

5. Click the = (equals) button.

6. Click **Mammal**.

7. Click **and**.

8. Click **Habitat**.

9. Click the = (equals) button.

10. Click **Tundra**.

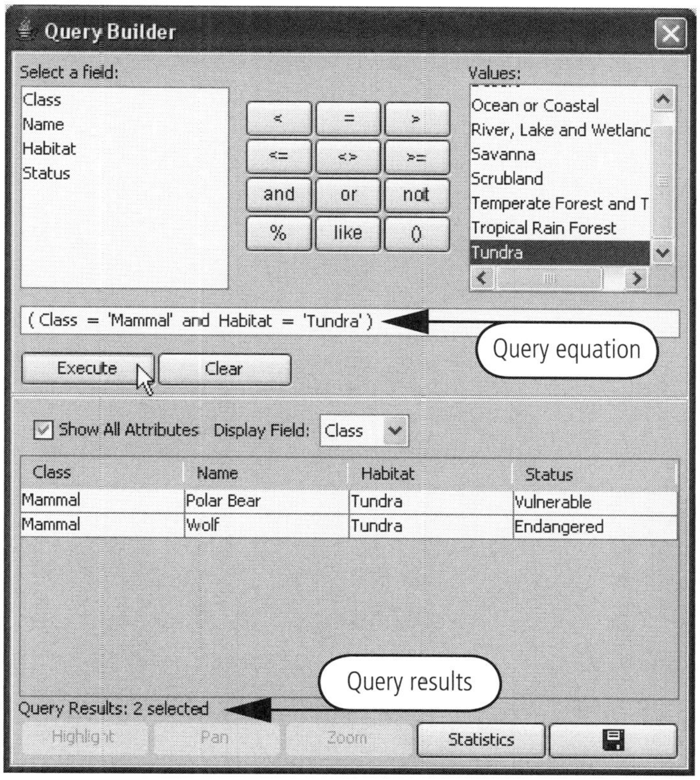

Your query should look like this:
(Class = 'Mammal' and Habitat = 'Tundra')
You are asking the GIS to show you all the animals that are mammals and that live in the tundra.

11. Click **Execute**.

The list of animals that meet these criteria will appear in the list at the bottom of the window.

Q15 **How many animals were selected?** ______________________________

Q16 **Write the names of all the animals under the Query answers column on your worksheet.**

Q17 **Have you ever seen one of these mammals in pictures or in real life? Write their names here.** ______________________________

____

____

____

Q18 **Write down some things you know about one of the animals.** ______________

____

____

____

12. Close the **Query Builder**. You will see the same animals in the list highlighted in green on the map.
13. Click the **Clear All Selections** button.

Step 7: Help save endangered species.

Thousands of animal species all over the world are considered endangered. There are several different levels of endangerment. Some animals have very low populations left and are considered endangered species because they are on the verge of disappearing from the earth (becoming extinct). Other animals are considered threatened species because they may become endangered in the future.

Zoos are often home to threatened and endangered species. Zoos are a way to help preserve their existence. Visitors and researchers appreciate being able to observe and study these animals. Many endangered animals are mammals that live in the African savanna. See if you can locate those animals in the zoo because you need to choose two of them to visit during the tour.

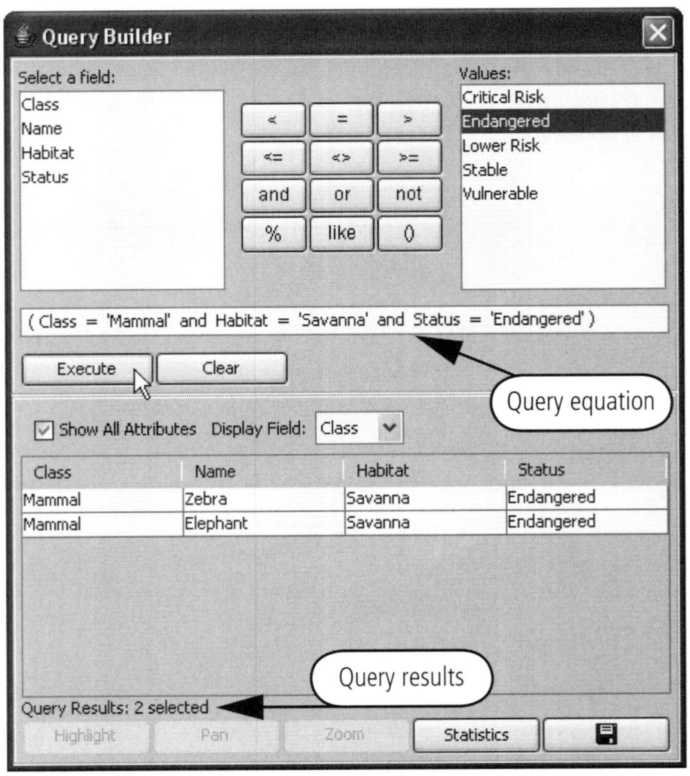

1. Make sure the **ZooAnimals** layer is still active. (Hint: It should be highlighted in dark blue in the table of contents.)
2. Click the **Query Builder** button. This opens the Query Builder window. You are once again going to build a complex query where you ask the GIS more than one question at a time.
3. Click **Clear** to erase the old query.
4. Under **Select a Field**, click **Class**.
5. Click the = (equals) button.
6. Click **Mammal**.
7. Click **and**.
8. Click **Habitat**.

9. Click the = (equals) button.
10. Click **Savanna**.
11. Click **and**.
12. Click **Status**.
13. Click =.
14. Click **Endangered**.

Your query should look like this:
(Class = 'Mammal' and Habitat = 'Savanna' and Status = 'Endangered').
You are asking the GIS to show you all the animals that are mammals and that live in the savanna and that are endangered. That's a pretty complicated question!

15. Click **Execute**.

The list of animals that are endangered mammals that live in the savanna will appear in the lower half of the window.

Q19 **How many animals were selected?** ______________________________

Q20 **Write the names of all the animals under the Query answers column on your worksheet.**

Q21 **Have you ever seen one of these mammals in pictures or in real life? Write their names here.** ______________________________

____

____

____

Q22 **Write down some things you know about one of the animals.** ______________

____

____

____

16. Close the **Query Builder**.

    You will see the same animals in the list highlighted in green on the map.

17. Click the **Clear All Selections** button.

Step 8: Decide which animals to visit.

Now that you have made a list of animals that meet the requirements you got from the zoo, select the animals you want to visit and in which order. Don't forget to keep lunch and shopping in mind too!

1. Right-click the **ZooAnimals** layer and choose **Properties**.
2. Click the **Labels** tab.
3. Under **Label features using**, choose **Name**.
4. Change the **Font** to **Arial**.
5. Place a check mark next to **Bold**.
6. Change the **size** to **12**.

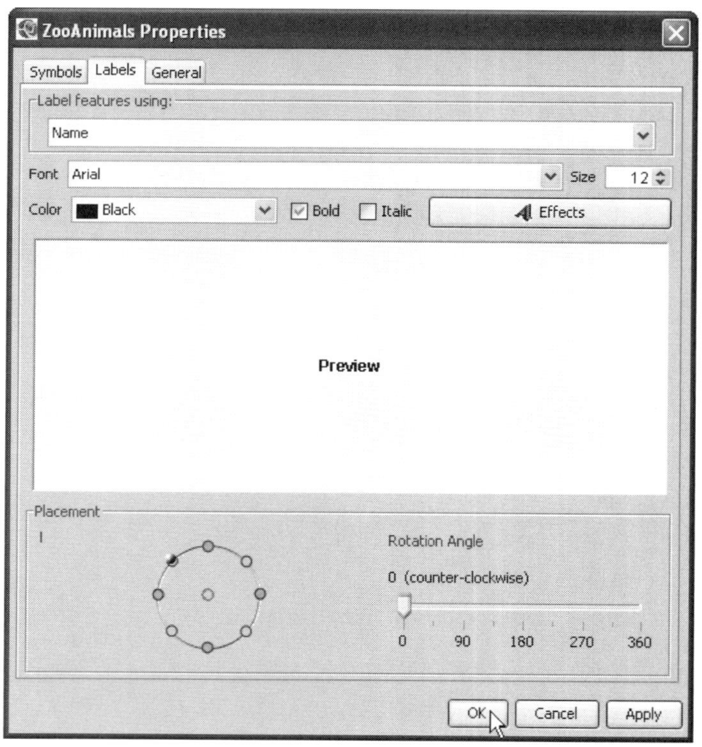

7. Click **OK**. Now all the animals are labeled by name.
8. Get your **worksheet 1** ready. Now you can decide which animals to include on your tour.

9. Look under the Query answers column. Choose one animal from each list and write it in the **Animal** column.

Example:

Class	Query answers	Animal (Write one animal from each query answer that you want to visit)	Order in tour
Reptile	1. Iguana 2. Anaconda 3. Snapping turtle 4. Rattlesnake 5. Crocodile	Snapping turtle	

M2 L2

10. Get out the map of the zoo that you created in the first activity (or use **worksheet 2**).

This map will look like the one on your computer screen. On the paper map you are going to draw the new tour route that visitors will take.

11. Look on the paper map and find the locations of all the animals you want to visit. These are the animals that you wrote in the **Animal** column on the worksheet.
12. Using the map, decide on the best order to visit the animals, stop for lunch, and shop in the gift shop. Make sure you go in a nice circle around the zoo. You don't want to cross back and forth.
13. Once you have decided the proper order, go to the **Order in tour** column on your **worksheet 1** and write the numbers 1 to 7 next to each stop in the order that you have chosen (you want to start at the Main Entrance).
14. Write the same numbers on your map in the correct locations.
15. On your paper map, draw a line that starts at the Main Entrance and joins all the points in order and ends at the Main Entrance.

Your map looks great! Now you can hand it out to visitors who request a custom tour of the zoo.

Step 9: Save your work and exit AEJEE.

1. Ask your teacher where and how to save your work.
2. Click the **File** menu and choose **Exit**.

Q23 From the tour you created, can you tell what additional animals will be seen along the way (besides the ones you selected)? List three of them here and two attributes (characteristics) about each animal.

Animal	Attribute 1	Attribute 2

Conclusion

A great deal of planning and research goes into creating a custom tour of a zoo or any other tourist attraction. You want to make sure that the visitors get the most information and best look at the zoo.

All the animals have different characteristics and come from unique places. Identifying the animals and their characteristics on a map can make it more interesting to study them and learn more about their lifestyles.

Name: ______________________________

Date: ______________________________

Module 2, Lesson 2

Touring a zoo

M2
L2

Worksheet 1:

This handout will help you organize and keep track of your answers in the lesson. Enter the answers to your queries in the spaces below. In some cases, there will be more animals than you need. Choose one animal from each query answer to use in your tour.

Class	Query answers	Animal (Write one animal from each query answer that you want to visit)	Order in tour
Reptile	1.		
	2.		
	3.		
	4.		
	5.		
Bird	1.		
	2.		
	3.		
	4.		

Continued on the next page.

Class	Query answers	Animal (Write one animal from each query answer that you want to visit)	Order in tour
Animals that live in the Scrubland	1.		
	2.		
	3.		
Mammals that live in the Tundra	1.		
	2.		
Mammals that live in the Savanna and are endangered	1.		
	2.		
You need stop for lunch at the restaurant			
You need to stop at the gift shop			

Worksheet 2: Touring the Zoo

Name: ______________________ Date: ______________

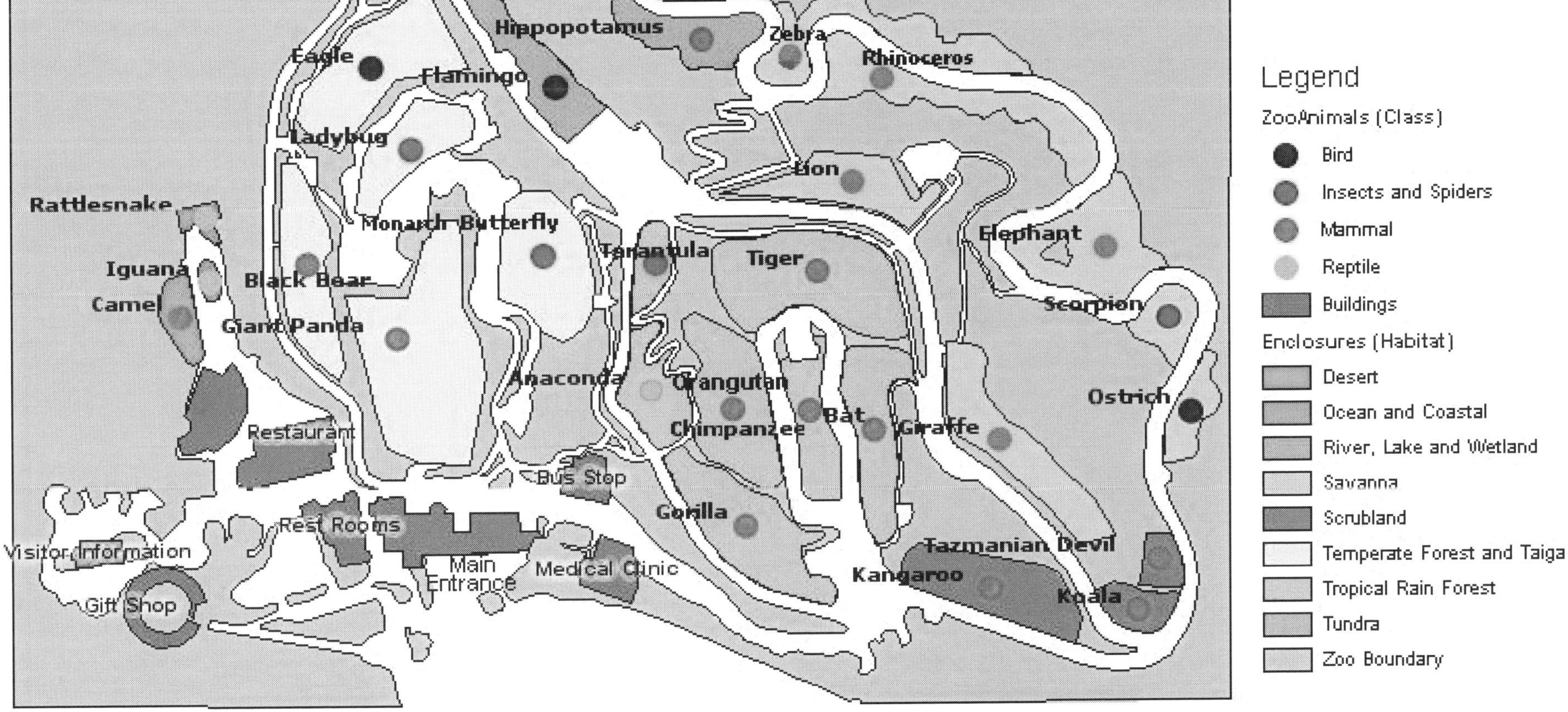

M2
L2

Name: ______________________________

Date: ______________________________

Module 2, Lesson 3

Animals around the world

The zoo in your community is so popular and successful that it has decided to expand. After careful research, zookeepers have decided to add an exotic animal to the zoo population. They are holding a contest for visitors to guess what the new animal will be. You will use skills you have learned in classification and analysis to find what part of the world the new animal is from and then identify it.

To help you get started, the zoo has provided a list of possible animals. A list of clues will help you choose the correct answers. You will combine information you have in multiple layers of maps to find your answer.

Step 1: Start AEJEE.

1. Ask your teacher how to start the AEJEE software.
2. Click the **Maximize** button at the top right of the window. Now the AEJEE window fills your screen.

Step 2: Open the project.

1. Click the **Open** button.
2. Navigate to **OurWorld1\Module2**.
3. Choose **MysteryAnimal.axl**.
4. Click **Open**.

   A map of the world appears on your screen with some other background information. A legend to the left contains a list of the layers you will be working with. A series of buttons and tools will help you perform GIS functions.

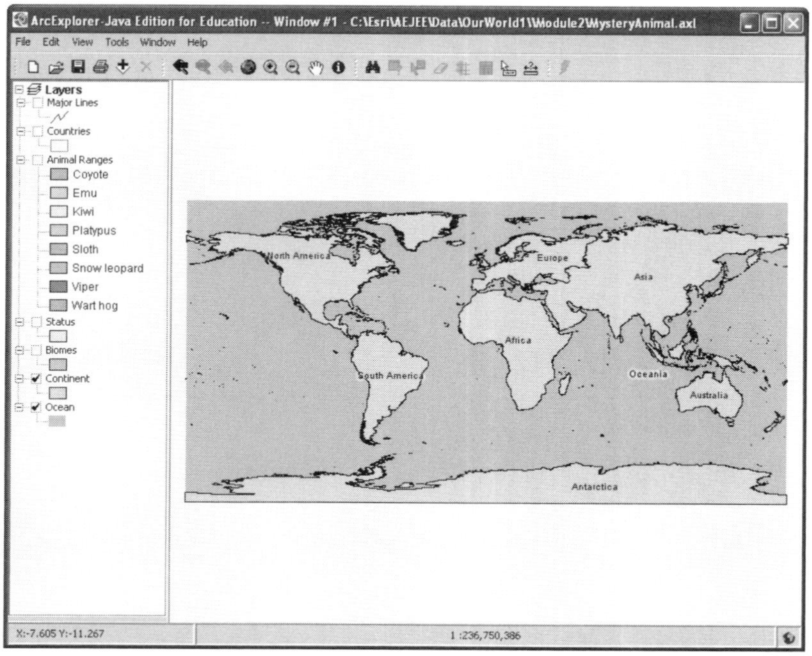

Step 3: Clue #1—The mystery animal lives in the Eastern Hemisphere.

The Eastern Hemisphere is the area of the world found east of the prime meridian and west of the International Date Line.

1. Turn on the **Animal Ranges** layer by placing a check mark next to the layer name.

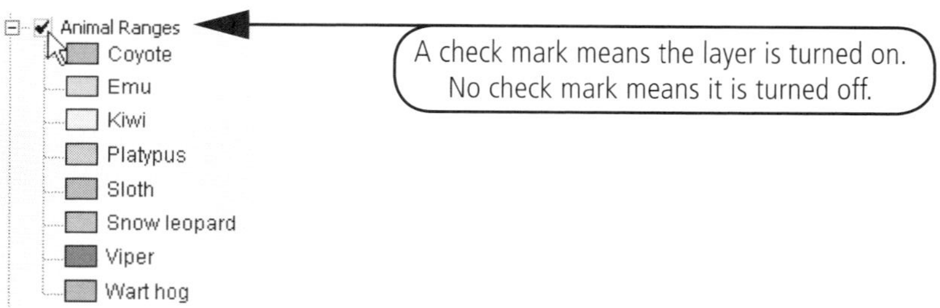

The legend lists the names of new animals the zoo is considering. The map shows where those animals live in the world. A different color represents each animal. The same list of animals appears in your "**Animal list" worksheet.**

2. Turn off the **Animal Ranges** layer by clicking the check mark next to the layer name.

The animal that you are looking for lives east of the prime meridian (0° longitude).

3. Turn on the **Major Lines** layer by clicking the box next to the layer name.

This is a layer showing the major lines of latitude in the world.

4. Locate the intersection of the prime meridian and the North Pole.

5. Click the **Zoom In** tool.

6. Click and hold your mouse where the prime meridian and the North Pole meet.

7. Drag a box around the entire Eastern Hemisphere and stop at the far southeast tip.

8. Release your mouse button.

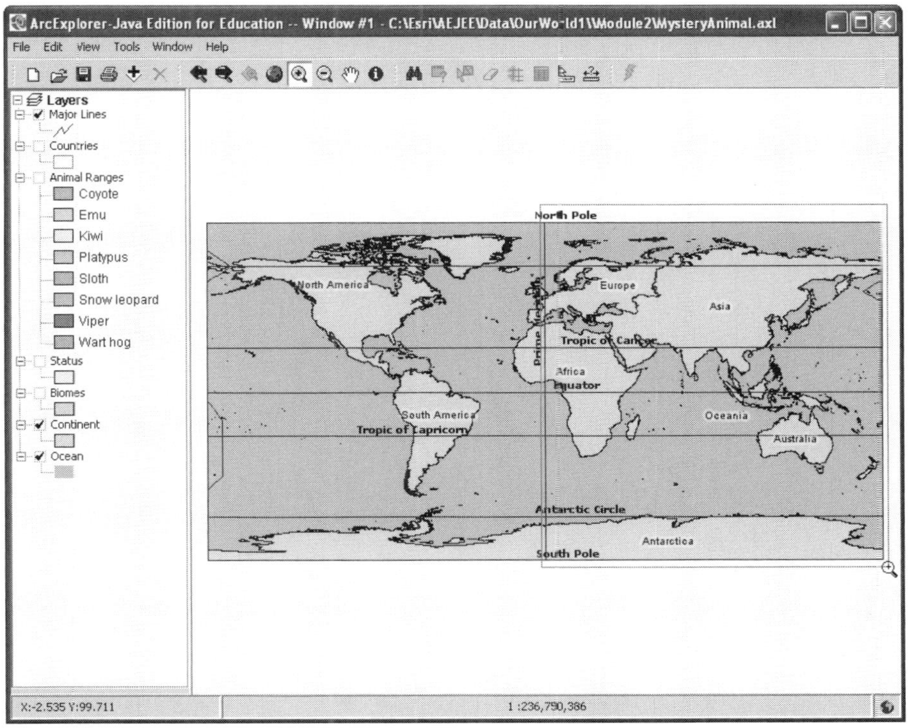

You will now have a closer look at the continents in the Eastern Hemisphere.

> **Note:** If you make a mistake, click on the **Zoom to Full Extent** button  and start over.

9. Turn on the **Animal Ranges** layer.

10. Click on the **Animal Ranges** layer name to highlight it. It will become active.

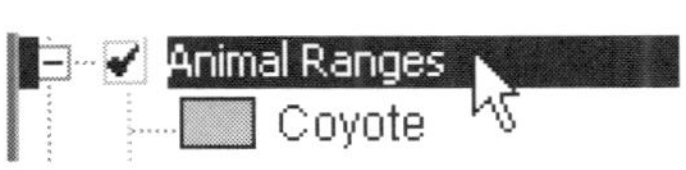

11. Click the **Identify** tool.

12. Click each animal range located in the Eastern Hemisphere. The name of the animal will appear in the Identify Results window.

Q1 **On your "Animal list" worksheet, place an X under the animals located in the Eastern Hemisphere (either part of or all of their range is in the hemisphere).**

____

13. Turn off the **Animal Ranges** layer.

Step 4: Clue #2—The mystery animal lives near the Tropic of Capricorn.

The **Tropic of Capricorn** is a major line of latitude located at approximately 23.5°S. This is the most southerly latitude where the sun can appear directly overhead at noon. This happens during the December solstice—about December 21—the official change from the spring season to the summer season (remember, the seasons are reversed in the Southern Hemisphere!).

Q2 **Look at your map. Is the Tropic of Capricorn located in the Northern Hemisphere (north of the equator) or in the Southern Hemisphere (south of the equator)?**

a. Northern Hemisphere

b. Southern Hemisphere

Q3 **Look at your map. What continents does the Tropic of Capricorn cross? Circle all that apply.**

a. Africa

b. Europe

c. North America

d. Australia

1. Click the **Zoom In** tool.
2. Click and hold your mouse at the intersection of the equator and the prime meridian.
3. Drag a box to the far southeast corner of your map.

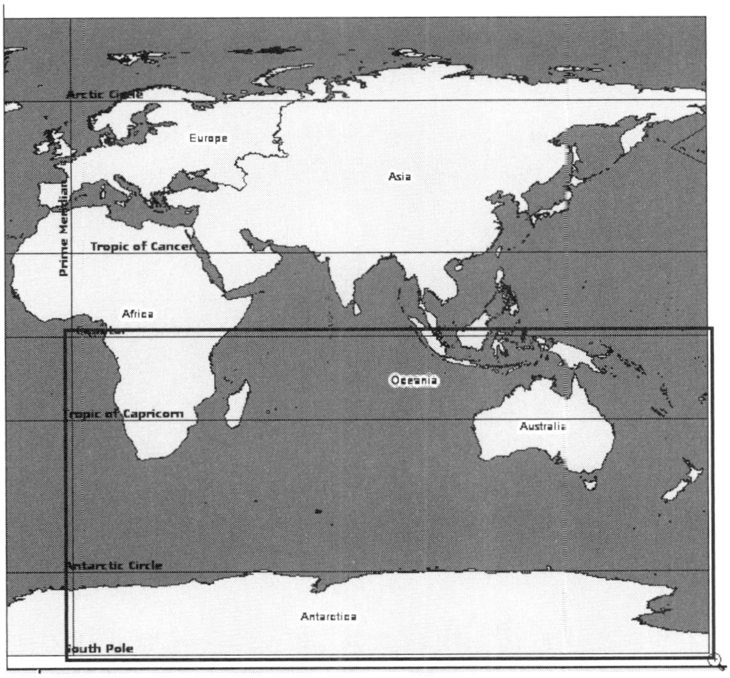

4. Release your mouse button.

You are now zoomed in to the Southern Hemisphere, east of the prime meridian.

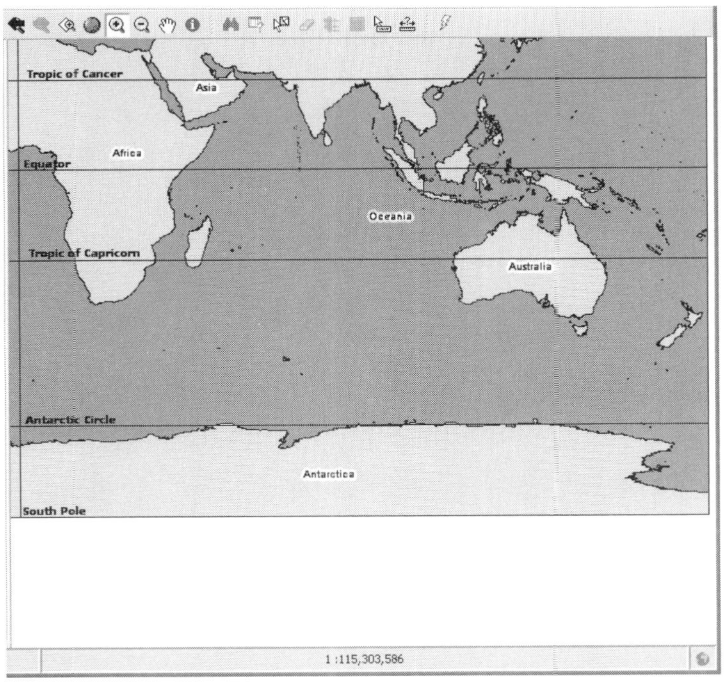

5. Turn on the **Animal Ranges** layer.
6. Make sure the **Animal Ranges** layer is still active. (Hint: It should be highlighted in dark blue in the legend.)
7. Click the **Identify** tool.

M2 L3

8. Click each animal range that is visible in the Southern Hemisphere.

Q4 **On your "Animal list" worksheet, place an X next to the animals that are located in the Southern Hemisphere (either part of or all of their range must be in the hemisphere).**

____

9. Turn off the **Animal Ranges** layer.

Step 5: Clue #3—The mystery animal lives in temperate mixed broadleaf forest areas.

A biome is an area that has similar animals, plants, and climate. A biome is a way of classifying the world into natural regions so that it's easier for people to study and research them.

Let's use the information you have to make a map showing the world's biomes.

1. Click the **Zoom to Full Extent** button. This will zoom the map out to the whole world so that you can see all the biomes.
2. Turn off the **Continent** layer.
3. Turn on the **Biomes** layer.

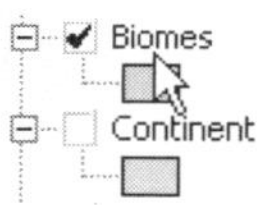

4. Right-click the **Biomes** layer and choose **Properties**.

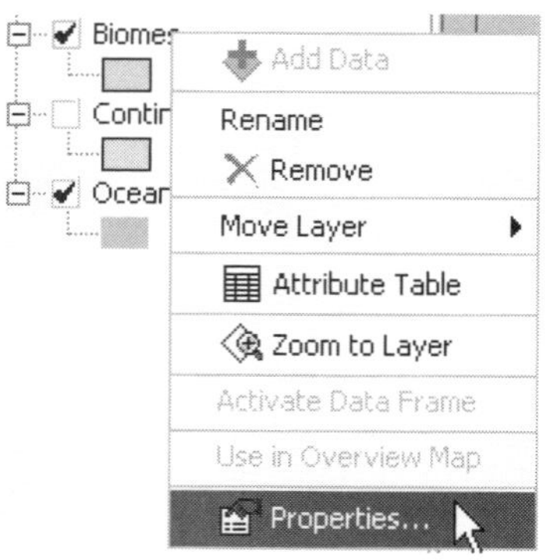

5. Under **Draw features using,** choose **Unique Symbols**.
6. Under **Field for values,** choose **BIOME_DESC**. (This field contains the descriptive name for each biome.)
7. Change the **Color scheme** to **Pastels**.
8. Click the box next to **Remove Outline**.
9. Click **OK**.

   Now each biome is identified with a different color.

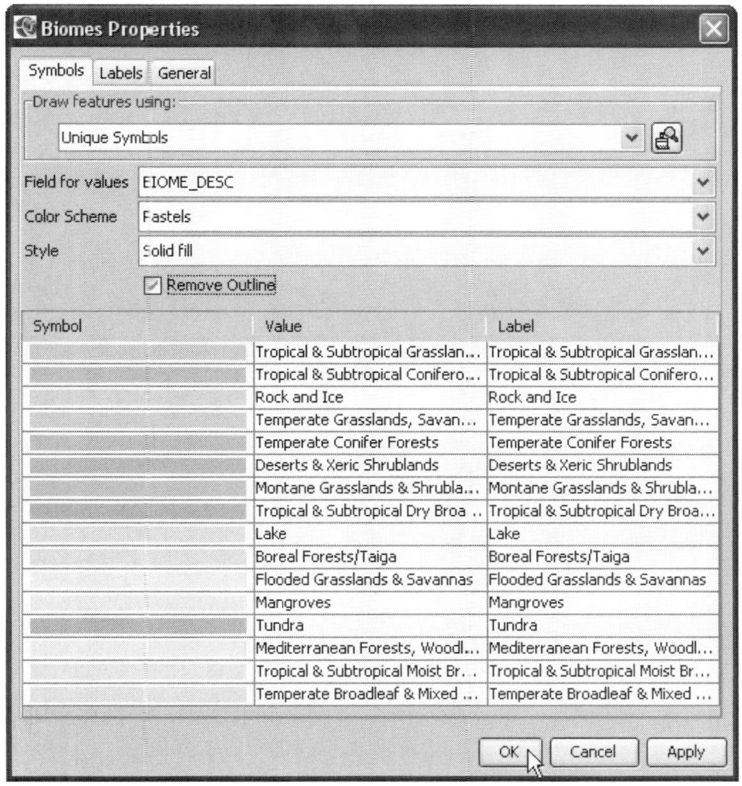

M2
L3

10. Right-click the **Biomes** layer and choose **Attribute Table**.

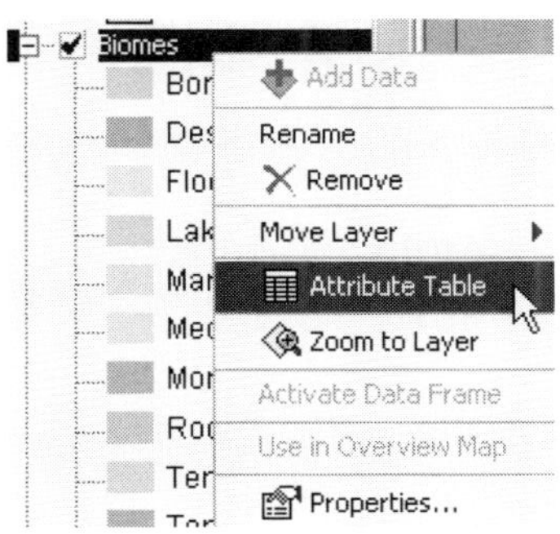

11. Locate the **BIOME_DESC** field heading.
12. Look down the list until you see **Temperate Broadleaf & Mixed Forests**.
13. Click the record (row) containing **Temperate Broadleaf & Mixed Forests**.

    It will become highlighted in blue in your table.

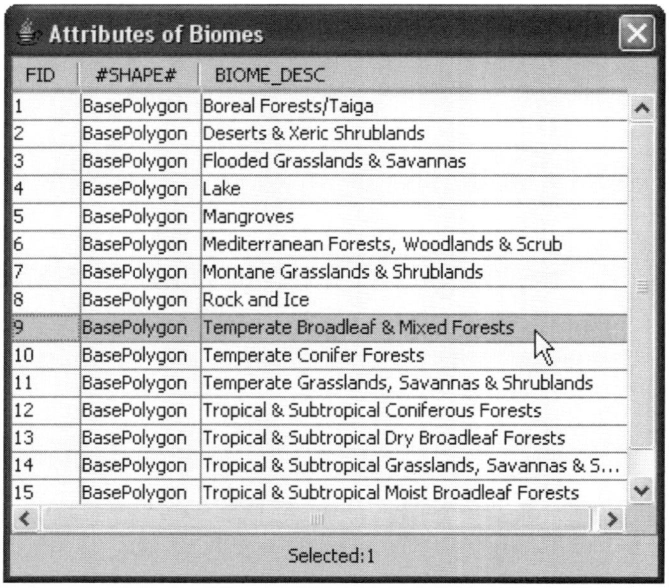

Attributes of Biomes

FID	#SHAPE#	BIOME_DESC
1	BasePolygon	Boreal Forests/Taiga
2	BasePolygon	Deserts & Xeric Shrublands
3	BasePolygon	Flooded Grasslands & Savannas
4	BasePolygon	Lake
5	BasePolygon	Mangroves
6	BasePolygon	Mediterranean Forests, Woodlands & Scrub
7	BasePolygon	Montane Grasslands & Shrublands
8	BasePolygon	Rock and Ice
9	BasePolygon	Temperate Broadleaf & Mixed Forests
10	BasePolygon	Temperate Conifer Forests
11	BasePolygon	Temperate Grasslands, Savannas & Shrublands
12	BasePolygon	Tropical & Subtropical Coniferous Forests
13	BasePolygon	Tropical & Subtropical Dry Broadleaf Forests
14	BasePolygon	Tropical & Subtropical Grasslands, Savannas & S...
15	BasePolygon	Tropical & Subtropical Moist Broadleaf Forests

Selected:1

14. Close the **Attribute Table**.

You will see all the temperate broadleaf and mixed forest regions outlined in bright yellow on your map. This shows the relationship between the attribute data and the map.

Temperate broadleaf and mixed forests are typically located between the tropics and the poles. The temperature changes in these areas are not extreme, but there are distinct warm and cool seasons. This area is home to a variety of animal species, including brown bears, owls, leopards, and ladybugs.

Now that you understand a bit about biomes, return to the southeastern part of the world and continue your search for the mystery zoo animal.

15. Click the **Zoom In** tool.
16. Find the intersection of the equator and the prime meridian.
17. Click, hold, and drag a box that starts there and covers the southeastern part of the world.

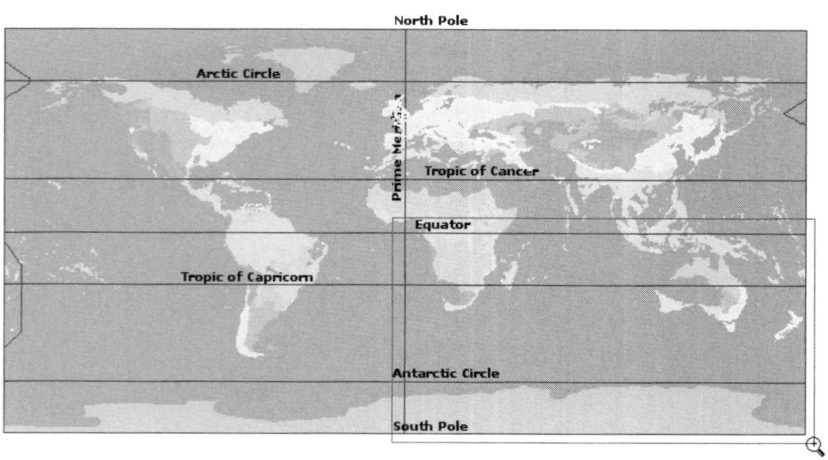

You are now zoomed back to the Southern Hemisphere.

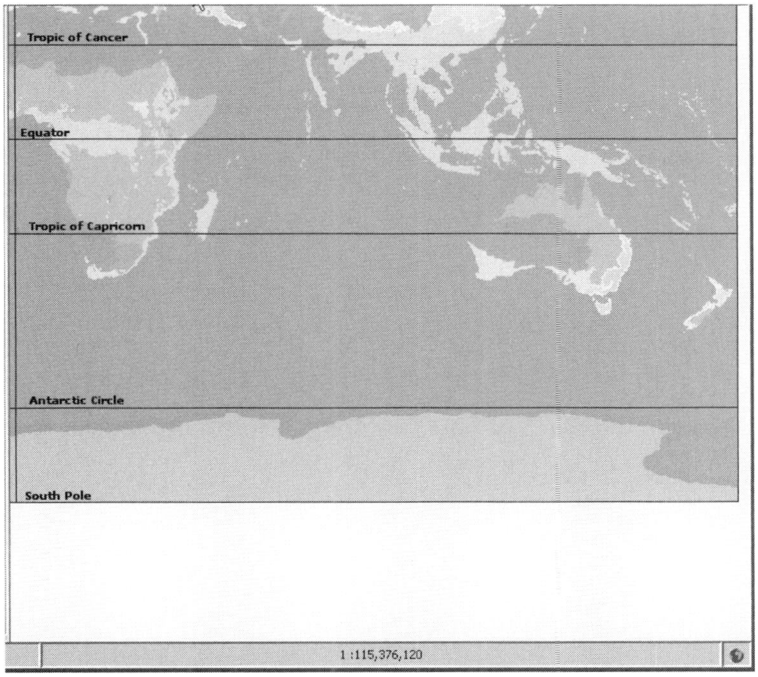

Remember, you are looking for temperate broadleaf and mixed forests in the southeastern portion of the world near the Tropic of Capricorn.

18. Turn on the **Countries** layer. The countries will appear with a black outline over the biomes.

19. Click the **MapTips** button.
20. Under **Layers**, choose **Countries**.
21. Under **Fields**, choose **CNTRY_NAME**.
22. Click **Set MapTips**.

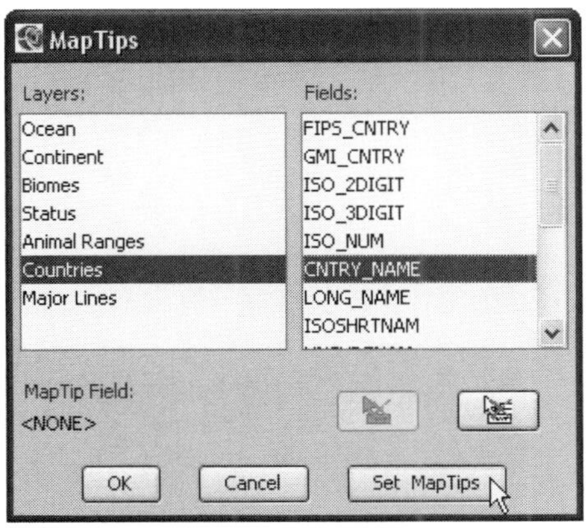

23. Click **OK**.
24. Place your pointer over the countries (don't click!), and the names of countries will appear.

M2
L3

Q5 **What two countries in this area have temperate broadleaf and mixed forest? (Hint: You are looking for the yellow highlighted areas south of the equator and east of the prime meridian.)**

1. ______________________________

2. ______________________________

25. Click the **Zoom In** tool.
26. Click, hold, and drag a box around these two countries.

> **Note:** If you make a mistake, click on the **Zoom to Previous Extent** button and start over.

27. Click the **Biomes** layer to make it active. It becomes highlighted.
28. Click the **Clear All Selections** button.

You have narrowed down the countries to Australia and New Zealand.

29. Turn off the **Biomes** layer.
30. Turn on the **Animal Ranges** layer.
31. Right-click the **Animal Ranges** layer and choose **Properties**.
32. Click the **Labels** tab.
33. Under **Label features using**, choose **COMM_NAME**.
34. Click **OK**.

Q6 **On your "Animal list" worksheet, place an X under the names of the animals found in these two countries.** ______________________________

Step 6: Clue #4—The mystery animal is widespread across its country but lives mostly in endangered or critical areas.

Now you will look at the conservation status of the ecoregions in each country.

Ecoregions are smaller than biomes. They each have their own plant, animal, and climate characteristics. The World Wildlife Fund has defined 825 ecoregions on earth. One type of information that the WWF records about ecoregions is their conservation status. There are three levels of classification:

- Stable or intact—Species and ecosystems are abundant and not at any significant risk.
- Vulnerable—Where the area is at *high* risk for the extinction of species and reduction of the ecosystems.
- Critical or endangered—Where species in this area are at a *very high* or *extremely high* risk of becoming extinct.

1. Turn off the **Animal Ranges** layer.
2. Turn on the **Status** layer by placing a check mark next to the layer name.

3. Right-click the **Status** layer and choose **Properties**.

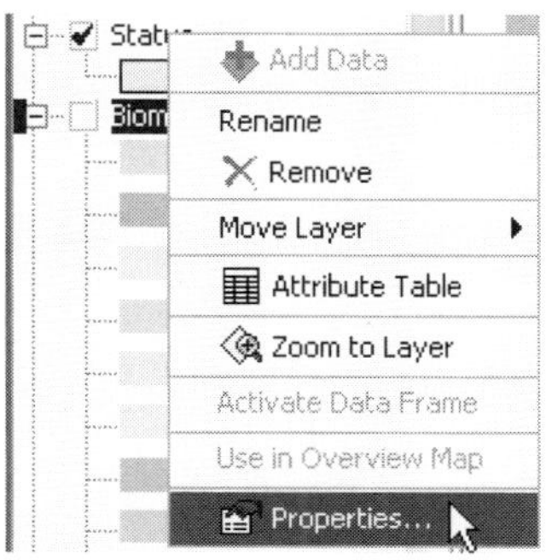

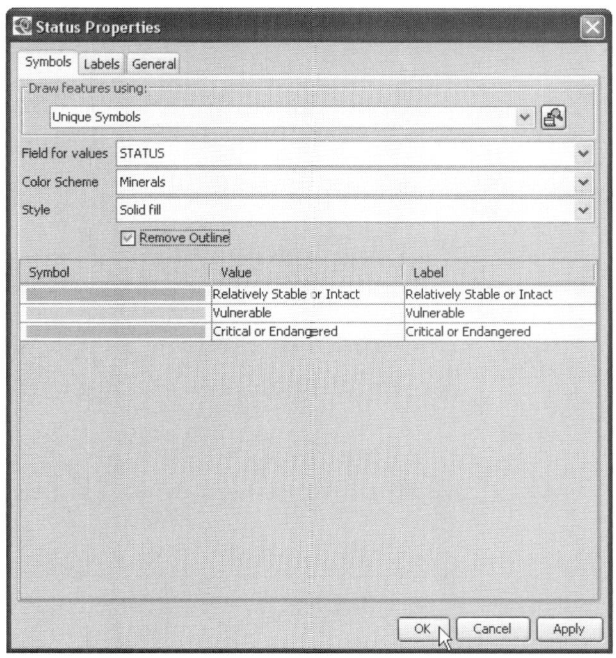

4. Under **Draw features using**, choose **Unique Symbols**.
5. Under **Field for values**, choose **STATUS**.
6. Change the **Color Scheme** to **Minerals**.
7. Place a check mark next to **Remove Outline**.
8. Click **OK**.

Q7 **Look at your legend. Which areas of Australia and New Zealand are critical or endangered?**

a. The coastal areas.

b. The central areas.

Q8 **Which country is mostly covered by critical or endangered areas?**

a. **Australia**

b. **New Zealand**

9. Turn on the **Animal Ranges** layer.
10. Click the **Animal Ranges** layer name to make it active.
11. Click the **Identify** tool.
12. Click the range for the animal that lives in the country you just identified.

Q9 **What is the name of the mystery animal?** ______________________________

Step 7: Placing the mystery animal.

Now that you have identified the mystery animal, you need to decide where it should live in the zoo.

Use the map of the zoo provided by your teacher (worksheet 2). Think about the following when you are choosing a new home for this animal:

1. Should you put the animal near other animals with similar characteristics or in a separate area?
2. What kind of habitat do you think it would like? Remember the biome that it was living in (go back to step 5).
3. Will it be seen on the tour that you created?

Q10 **Draw the new enclosure on your map—be sure to label the enclosure with the name of the zoo's newest resident.**

Q11 **Write a brief paragraph explaining the location you chose for the animal.**

____

____

Step 8: Save your work and exit AEJEE.

1. Ask your teacher where and how to save your work.
2. Click the **File** menu and choose **Exit**.

M2
L3

Conclusion

The kiwi is a small flightless bird that lives only in New Zealand. New Zealand has more flightless birds than any other country. Some people believe the birds there didn't need to fly because they didn't have many enemies to fly away from.

The kiwi bird is about the size of a chicken. Kiwis usually mate for life and typically lay only one egg at a time. The reason for this is that a kiwi egg can be up to 10 times larger than a chicken egg! Kiwis have an excellent sense of smell and are the only bird with nostrils at the end of its beak!

Unfortunately, the kiwi population in New Zealand faces threats from an increase in predators, human influence, and loss of habitat. Numerous organizations aim to help the kiwi begin to thrive again in New Zealand.

Name: ______________________________

Date: ______________________________

Module 2, Lesson 3

Animals around the world

M2 L3

Worksheet 1: Animal list

Class	Viper	Wart hog	Sloth	Emu
Lives in the Eastern Hemisphere?				
Lives in the Southern Hemisphere?				
Lives in the two countries found in step 5?				

Class	Coyote	Kiwi	Platypus	Snow Leopard
Lives in the Eastern Hemisphere?				
Lives in the Southern Hemisphere?				
Lives in the two countries found in step 5?				

Worksheet 2: Touring the Zoo

Name: ______________________ Date: ____________

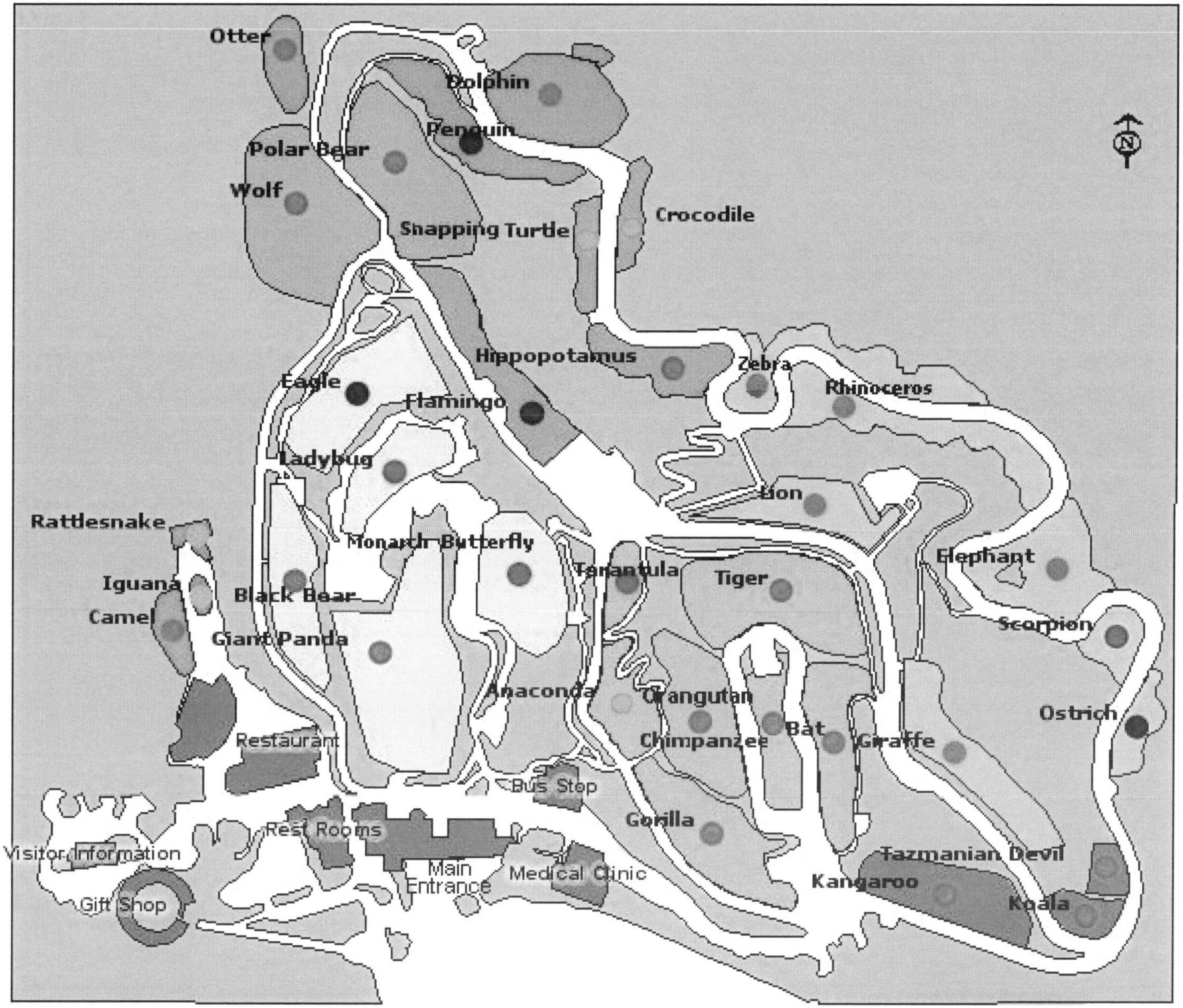

M2 L3

Name: ______________________________

Date: ______________________________

Module 3, Lesson 1

Early settlement patterns of the United States

Before Europeans came to America, Native Americans had been living on the land for thousands of years.

Starting in the early 1600s, many people from Europe started coming to America to settle the same lands where millions of Native Americans had lived, hunted, fished, farmed, and traveled. The first European settlements were along the Atlantic Coast.

M3 L1

In the first 150 years after Europeans started coming to America (until about 1765), European settlement spread from the Atlantic Coast to the Appalachian Mountains. In the next one hundred years, settlement advanced all the way to the Pacific Ocean. As European settlements spread westward, Native American tribes were killed or driven off their lands and forced to move westward onto reservations.

This lesson is about the movement of European peoples from the settled regions of the eastern United States to lands farther west. As these people moved west, their settlements formed different patterns on the land. Sometimes the settlements formed a regular pattern; other times, the settlements were irregular, or scattered. Sometimes the settlements were close to transportation routes, such as rivers; other times, they were located close to each other, forming clusters.

Let's explore the historic settlement patterns we can still see among the cities of today.

Step 1: Start AEJEE.

1. Ask your teacher how to start the AEJEE software.
2. Click the **Maximize** button at the top of the AEJEE window. Now the AEJEE window fills your screen.

Step 2: Open the project.

1. Click the **Open** button.
2. Navigate to your **OurWorld1\Module3** folder.
3. Choose **SettlementPatterns.axl**.
4. Click **Open**.

When the project opens, you see a map of the United States on your screen. (You don't see Alaska and Hawaii in this view.)

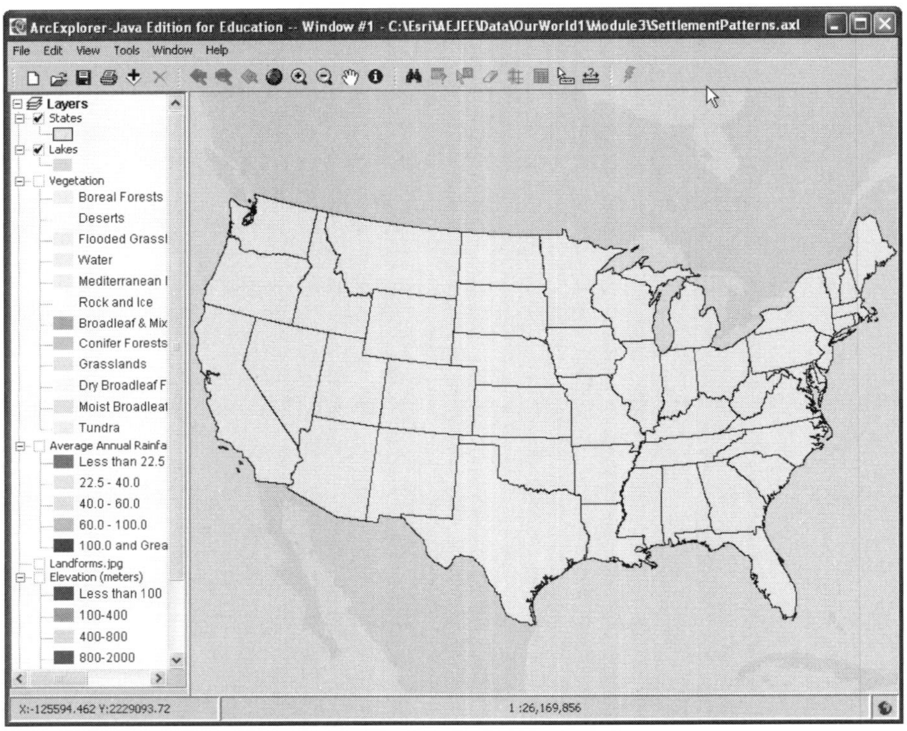

On the left side of the map, you see a list of layers. Layers are used to show geographic data on a GIS map. Each layer has a name and a legend. On the top, you see buttons and tools that you will use during this activity.

Take a minute to label the states.

5. Click the **States** layer name to highlight it. It becomes active.

Making a layer active tells the GIS that this is the layer you want to work with.

6. Right-click the **States** layer name and choose **Properties**.

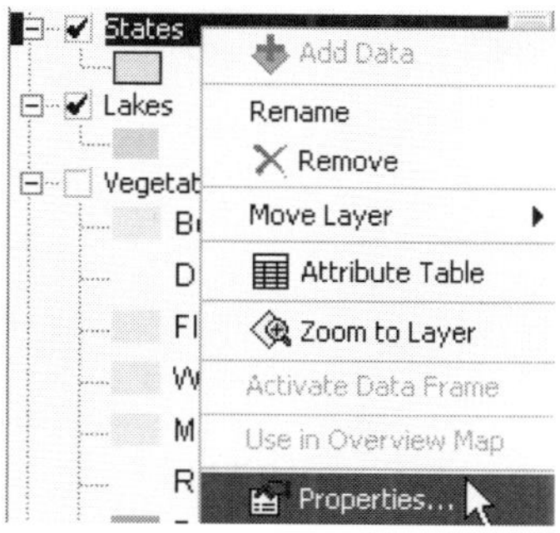

7. Click the **Labels** tab.
8. Under **Label features using**, choose **STATE_ABBR**. This field contains the abbreviations for the state names (e.g., California is CA)
9. Click **Effects**.
10. Click the box next to **Glow**.
11. Change the Glow color from Yellow to **Cyan**.
12. Click **OK**.
13. Click **OK** again.

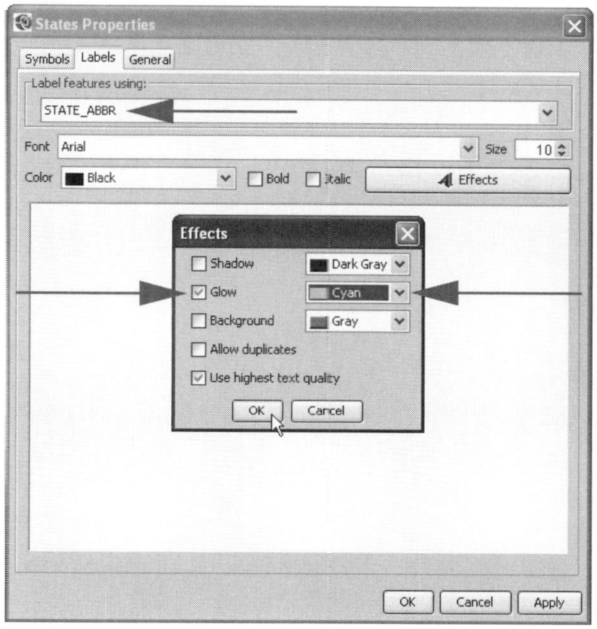

Now the states are labeled with their abbreviations, which have a glow around them. In order to see their full names, you will turn on MapTips.

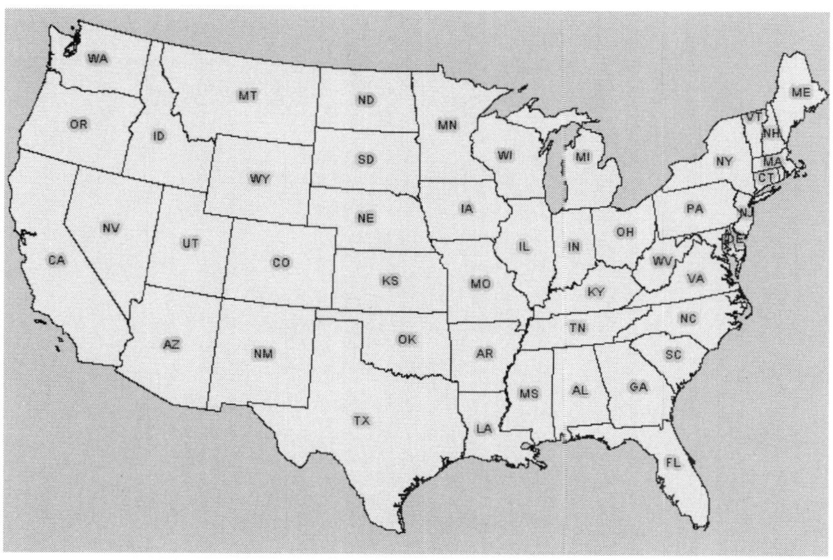

14. Click the **MapTips** button. The MapTips window opens.
15. Under **Layers**, click **States**.
16. Under **Fields**, click **STATE_NAME**.
17. Click **Set MapTips**.
18. Click **OK**.
19. Now when you hold your cursor over a state, the full name of the state appears.

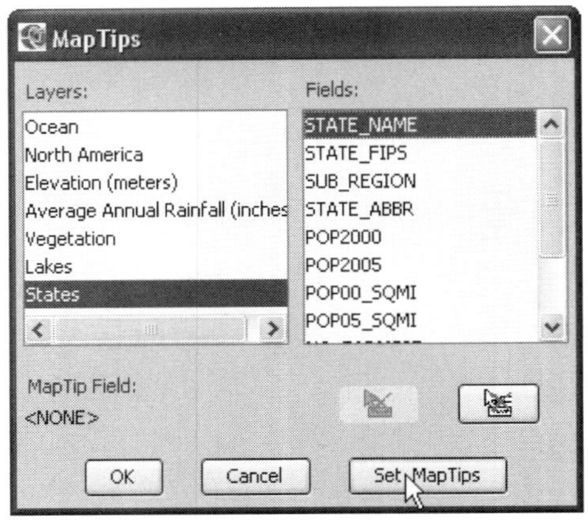

Q1 **Look at the sizes of the states. What pattern do you see when you move from east to west? (Circle all the correct answers.)**

a. Eastern states are smaller.

b. Eastern states are larger.

c. Western states are smaller.

d. Western states are larger.

Step 3: Settlements started along the East Coast.

The first European settlements developed along the East Coast. The British established their first settlement at Jamestown, Virginia, in 1607. Settlers established more towns in Plymouth and Boston, Massachusetts, and elsewhere between 1620 and 1630.

You will add cities to the map. Even though the cities you add are modern cities, they still reflect the settlement patterns that began with the early European settlements.

1. Click the **Add Data** button.
2. Navigate to your **OurWorld1\Module3\Data** folder.
3. Click **Cities.shp**.
4. Click **OK**.

Cities of the United States appear on the map. You'll change their color.

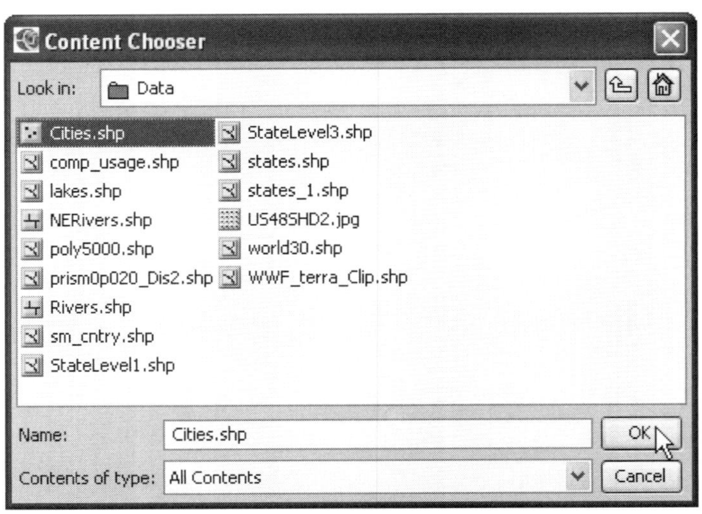

5. Right-click the **Cities** layer name and choose **Properties**.

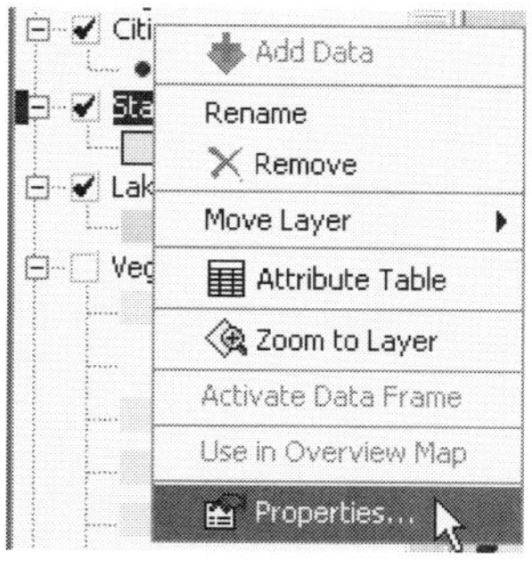

6. Next to **Color,** click the down arrow and choose **Green**.
7. Click the check mark next to **Remove Outline** to uncheck it.

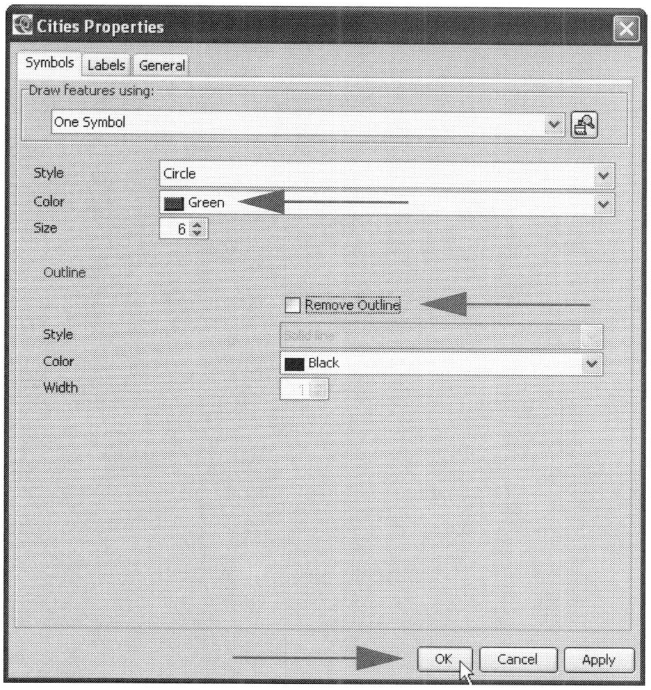

8. Click **OK**.

Now all the cities appear as green dots.

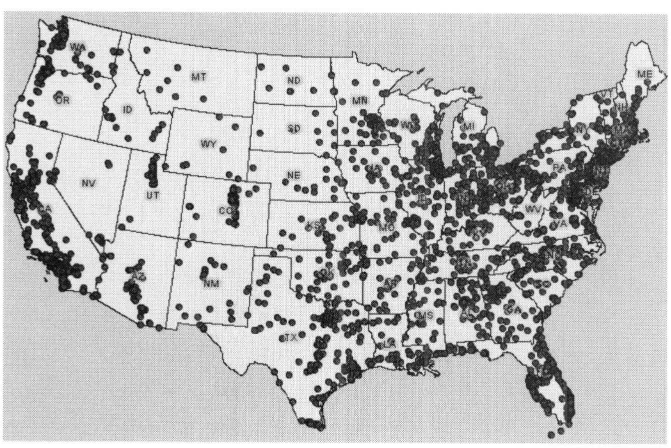

M3
L1

Q2 **Which half of the country has more cities?**

a. **Eastern half**

b. **Western half**

Q3 **Look at the states along the East Coast. List three of the states that have cities along their eastern shore. (Hint: Use MapTips)**

1. ______________________________

2. ______________________________

3. ______________________________

Step 4: When European settlers moved west from the coast, they often followed waterways.

People used lakes and rivers to travel to new places to settle. People also used lakes and rivers to carry goods and travel from areas far from the ocean back to the shore, where most trading with Europe took place. This was especially true in the northeastern states.

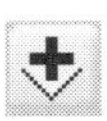

1. Click the **Add Data** button.

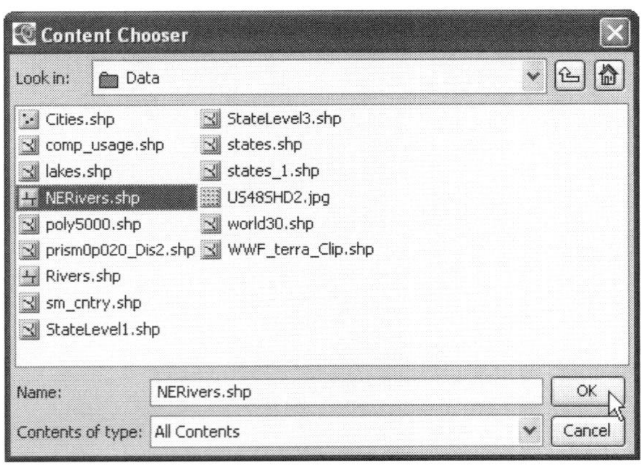

2. Navigate to the **OurWorld1\Module3\Data** folder.
3. Click **NERivers.shp**.
4. Click **OK**.

   Rivers of the northeast appear on the map. You will zoom in for a closer look.

5. Click the **NERivers** layer name to highlight it. It becomes active.

6. Click the **Zoom to Active Layer** button.

   The map zooms to the northeast part of the country. The rivers are hard to see, so you will change their symbol.

7. Right-click the **NERivers** layer name and choose **Properties**.
8. Click the down arrow next to **Color** and choose **Blue**.
9. Increase the **Size** to **3**.

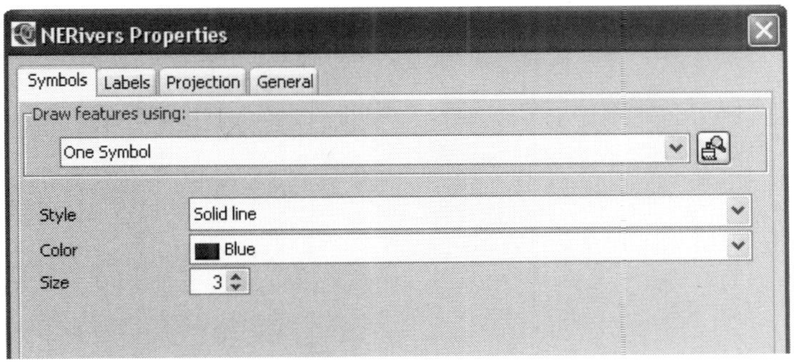

10. Click the **Labels** tab.
11. Under **Label features using** choose **NAME**.
12. For **Color**, choose **Blue**.
13. Increase the **Size** to **12**.
14. Click the box next to **Bold**.
15. Click the box next to **Italic**.
16. Click the **Effects** button.
17. In the Effects window, click the box next to **Background**.
18. For the Background **Color**, choose **White**.
19. Click **OK**.
20. Click **OK** again.

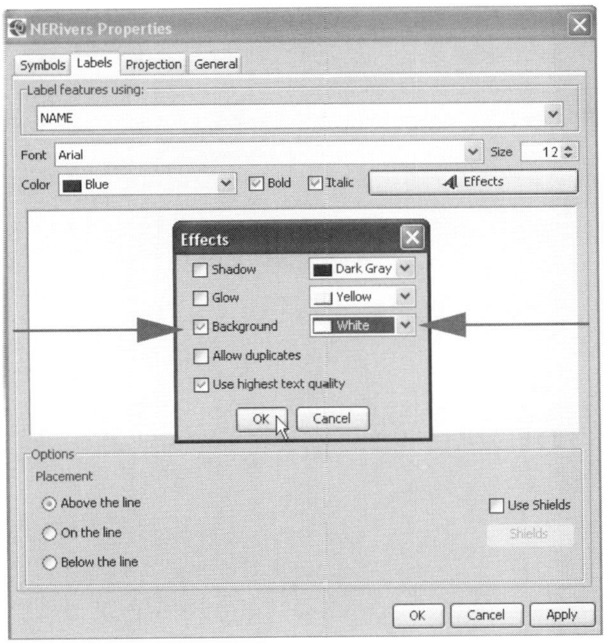

Now the northeast rivers appear as thick blue lines with blue labels on white backgrounds.

M3
L1

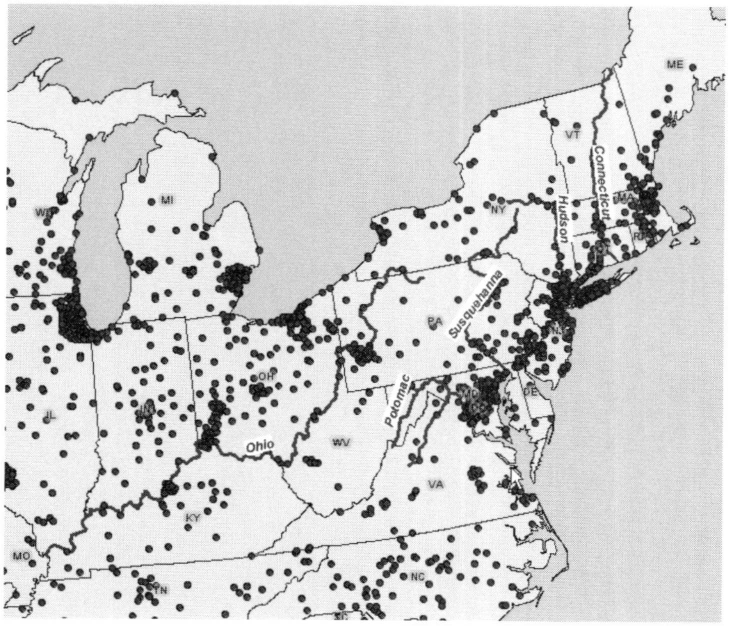

The British people used the waterways as highways, just like the Native Americans did. Some settled along the coast of Chesapeake Bay. It was here that the Native Americans introduced the settlers to tobacco.

21. To find Chesapeake Bay, look at your map and follow the clues below.
22. Clue #1: Find the Susquehanna River. (Hint: It has a blue label.)
23. Clue #2: Find the place where the Susquehanna River empties into the ocean.

Now you have found Chesapeake Bay.

Q4 **Are there cities along the coast of Chesapeake Bay? (Circle the correct answer.)**

a. Yes

b. No

Q5 **Most of these cities are in what state? (Hint: Use MapTips.)** ____________________

The Dutch settled along the Hudson River. Henry Hudson, an Englishman working for a Dutch trading company, was the first European to discover the Hudson River Valley. He was sailing along America's north Atlantic Coast when he found it by accident in 1609. He entered what is now New York Bay and sailed up the river that was later named for him. His explorations led to the area being settled by the Dutch between 1609 and 1664. The Dutch settlement was first known as New Amsterdam, but was later renamed by the British to New York.

24. Turn off the **Cities** layer by clicking the check mark next to its name.

25. Look at the map and find the Hudson River. (Hint: It has a blue label.)
26. Find the place where the Hudson River empties into the ocean.
27. Turn the **Cities** layer back on.

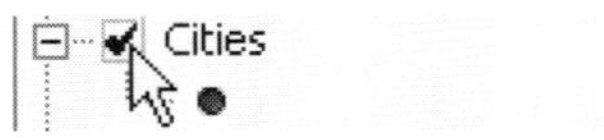

Q6 **Are cities found along the Hudson River where it empties into the ocean? (Circle the correct answer.)**

a. Yes

b. No

Q7 **Most of these cities are in what state? (Hint: Use MapTips.)** ____________

Step 5: By about 1750, European settlements had moved west as far as the Appalachian Mountains.

The early European settlers wanted to stay close to each other and to their home countries across the Atlantic. The rugged Appalachian Mountains discouraged them from moving farther west for many years because it was difficult to cross, with few openings and few trails for wagons.

1. Right-click the **States** layer name and choose **Properties**.
2. For **Style**, choose **Transparent fill**.

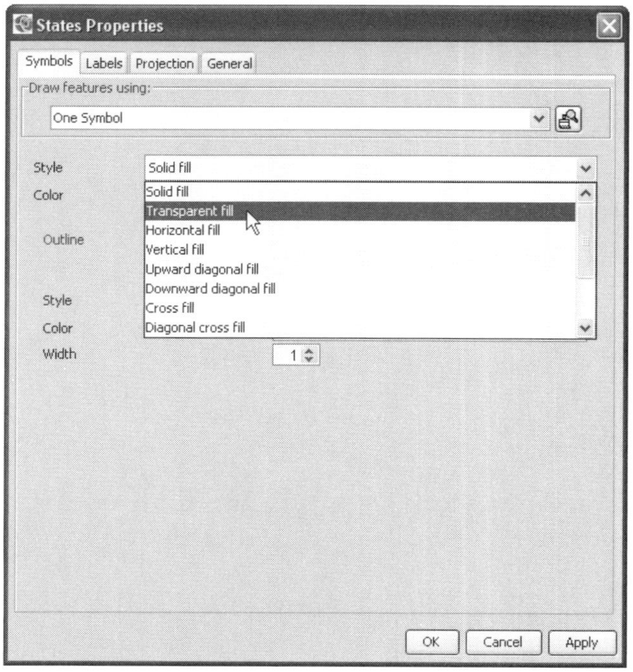

3. Click **OK**.

   Now you see just the outlines of the states with their labels.

4. Turn on the **Elevation (meters)** layer by clicking the box next to the Elevation layer name. (Be patient while this layer draws.)

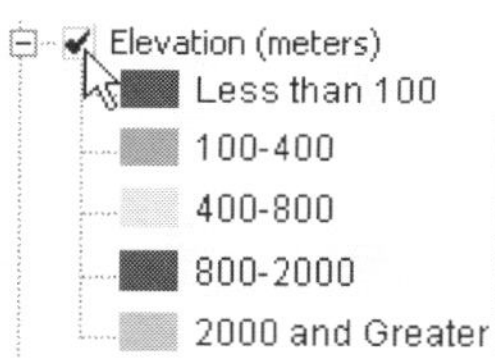

The elevation layer shows high areas (mountains) and low areas (plains).

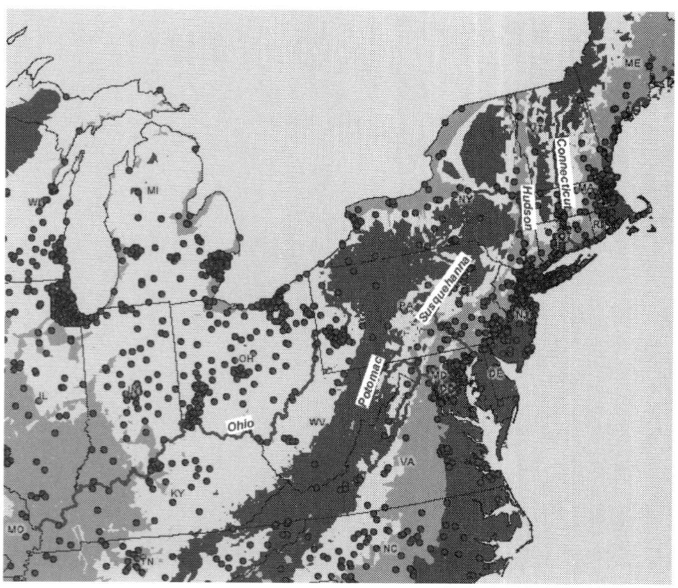

Q8 **Look at the map and the Elevation legend. Which color do you think represents the Appalachian Mountains? (Circle the correct answer.)**

a. **Dark green**

b. **Light green**

c. **Light brown**

d. **Dark brown**

e. **Pink**

Q9 **How high are the Appalachian Mountains in meters? (Circle the correct answer.)**

a. **Less than 100**

b. **100–400**

c. **400–800**

d. **800–2,000**

Hint: A meter is a little more than 3 feet (3.28 feet), so if you multiply the number of meters by 3, you will know approximately how high the mountains are in feet.

5. To see the mountains better, turn on the **Landforms.jpg** layer by clicking the box next to its name.

   This layer is an image (or picture) that shows the mountains and plains. It is partly see-through, so you can see the elevation layer at the same time. Now you can really see where the mountains are. You'll zoom out for a better view.

6. Click the **Zoom Out** tool.
7. Click the word **Ohio** near the center of the map. (This is the blue label for the Ohio River.)

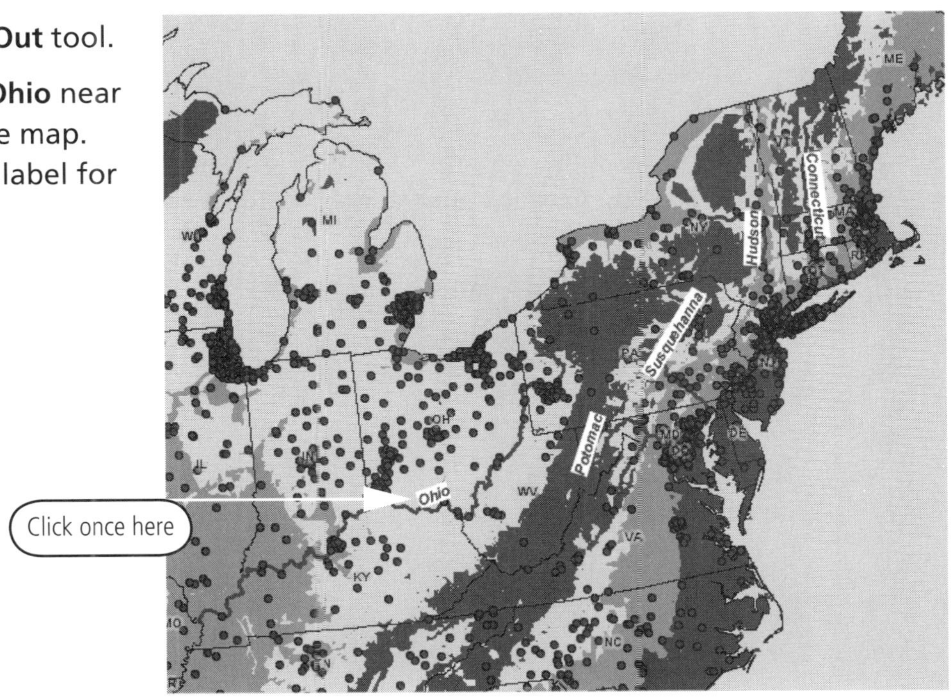

The map zooms out so you can see more of the Appalachian Mountains.

> **Note:** If you make a mistake, click the **NERivers** layer name to highlight it and click the **Zoom to Active Layer** button. Then click the **Zoom Out** tool and try again.

Step 6: When Europeans finally moved west of the Appalachian Mountains in about 1770, they quickly settled along waterways, such as the Ohio River.

Native Americans who lived west of the Appalachians wanted to keep their land and their way of life. They fought a bloody war with the British. Then the British government passed a law banning European settlers from moving west of the Appalachians.

However, this law was almost impossible to enforce because the British government was so far away across the Atlantic Ocean. In the 1770s, Richard Henderson, who hoped to make money in the land business, got permission from the local Native Americans (Cherokees) to settle in Kentucky. He hired the famous frontiersman Daniel Boone to lead families across the Appalachians at a place called the Cumberland Gap. They founded the town of Boonesboro in central Kentucky.

1. Turn off the **Landforms.jpg** layer by clicking the check mark next to its name.

2. Click the **Previous Extent** button to return to a closer view.
3. Look at the Ohio River on the map.

Q10 **Which of these patterns most closely matches the pattern of cities along the Ohio River? (Circle the correct answer.)**

a. Regular

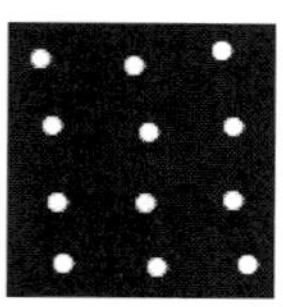

b. Clustered

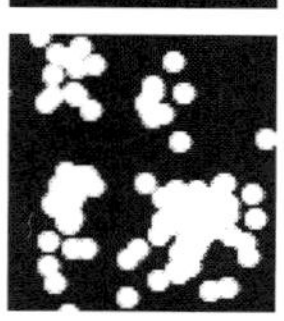

c. Random

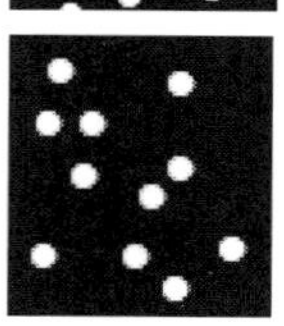

Step 7: Settlement continued to move west as far as Kansas and Nebraska.

After the Americans defeated the British in the American Revolution (1775–1783), the U.S. government encouraged people to move west by selling them land. The government paid war veterans by giving them free land.

European settlers brought a different kind of agriculture to the land. While Native Americans had grown several crops on a single piece of land, European settlers used one piece of land to grow a single crop. They also tried to grow more than they could eat, so they could sell the excess to people in the east.

The new approach to farming exhausted the soil more quickly than Native American farming, and it created a demand for more and more land. This helped push settlement farther westward.

1. Click the **Zoom Out** tool.
2. Place your mouse over the western end of the Ohio River and click once.

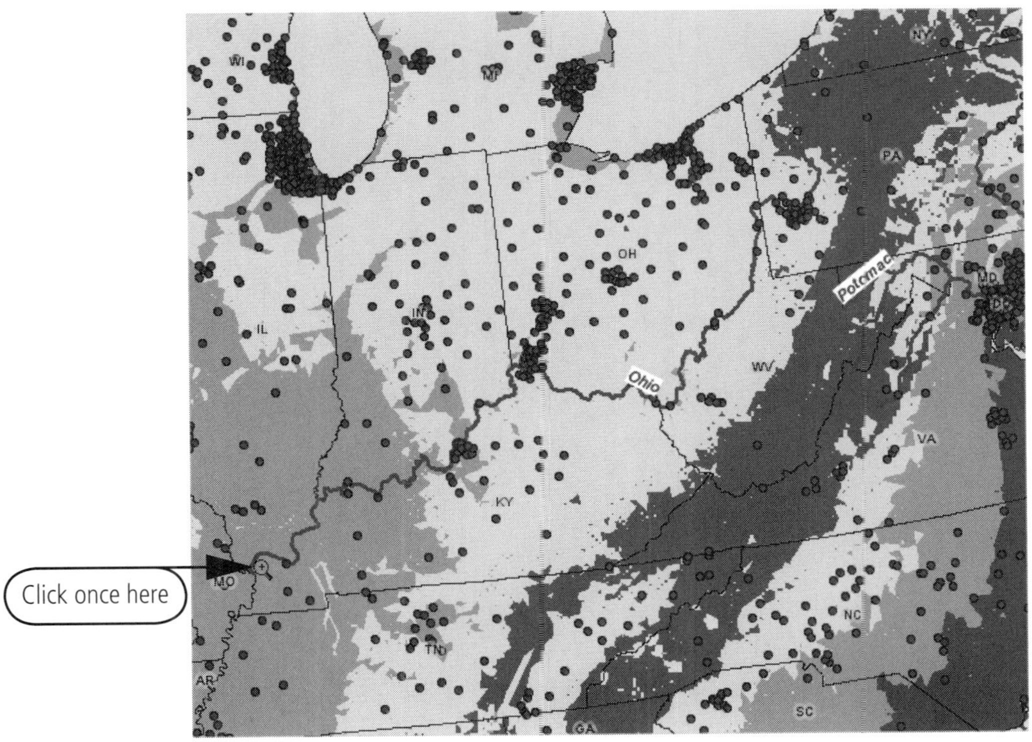

Note: If you make a mistake, click the **NERivers** layer to highlight it, and click the **Zoom to Active Layer** button. Then click the **Zoom Out** tool, and try again.

Q11 **Look at the map and the Elevation legend. What happens to elevation as you move west into Kansas, Nebraska, and South Dakota?**

a. **Elevation increases (elevation is higher)**

b. **Elevation decreases (elevation is lower)**

c. **Elevation stays the same**

Q12 **Look at the elevation layer on the map and compare it with the cities layer. Can you explain how elevation affects where people live?**

____

____

3. Turn off the **Elevation** layer by clicking the check mark next to its name.
4. Turn on the **Average Annual Rainfall** layer. (Be patient while this layer draws.)

   This layer shows the average amount of rain (in inches) that falls in a year.

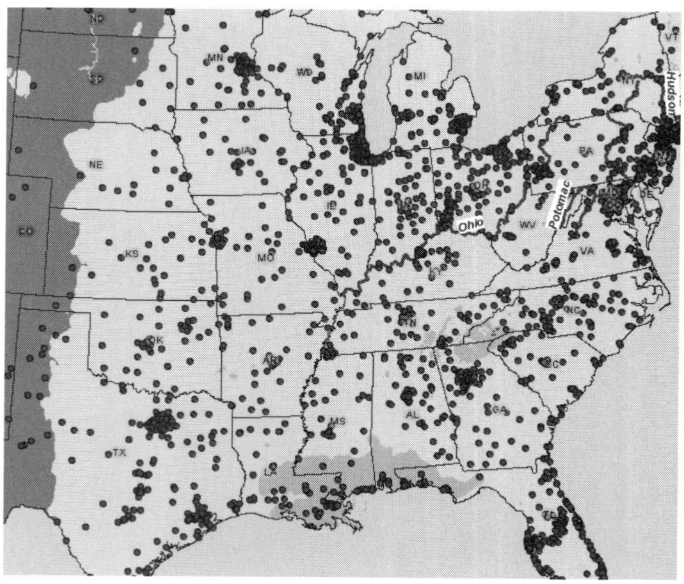

Q13 **Look at the legend for the Average Annual Rainfall layer in the table of contents. Which color best represents the amount of rainfall in the states of Oklahoma, Kansas, and Nebraska? (Circle the correct answer.)**

a. **Red**

b. **Orange**

c. **Green**

d. **Blue**

Q14 **How much average annual rainfall does this color represent? (Circle the correct answer.)**

a. **Less than 22.5 inches**

b. **22.5–40 inches**

c. **40–60 inches**

Q15 **Look at the rainfall layer and compare it with the cities layer. Can you explain how rainfall affects where people live?**

____

____

____

____

M3
L1

Step 8: By the 1840s, settlement had moved west of the Mississippi River onto grasslands in the center (interior) of the country.

1. Turn off **Average Annual Rainfall** by clicking the check mark next to its name.
2. Turn on the **Vegetation** layer.

   The Vegetation layer shows the type of plant growth in an area. (Example: Grasslands)

3. Click the **View** menu and choose **Pan Panel**.

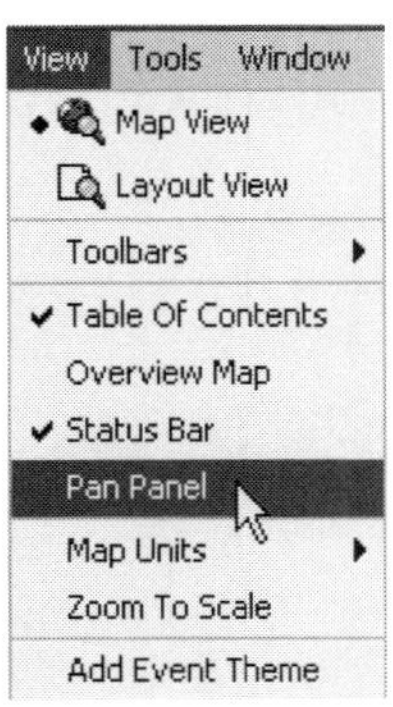

   A frame appears around your map. Each side of the frame has a small white arrow.

4. Click **once** on the small white arrow on the left side of the frame to **Pan West**.

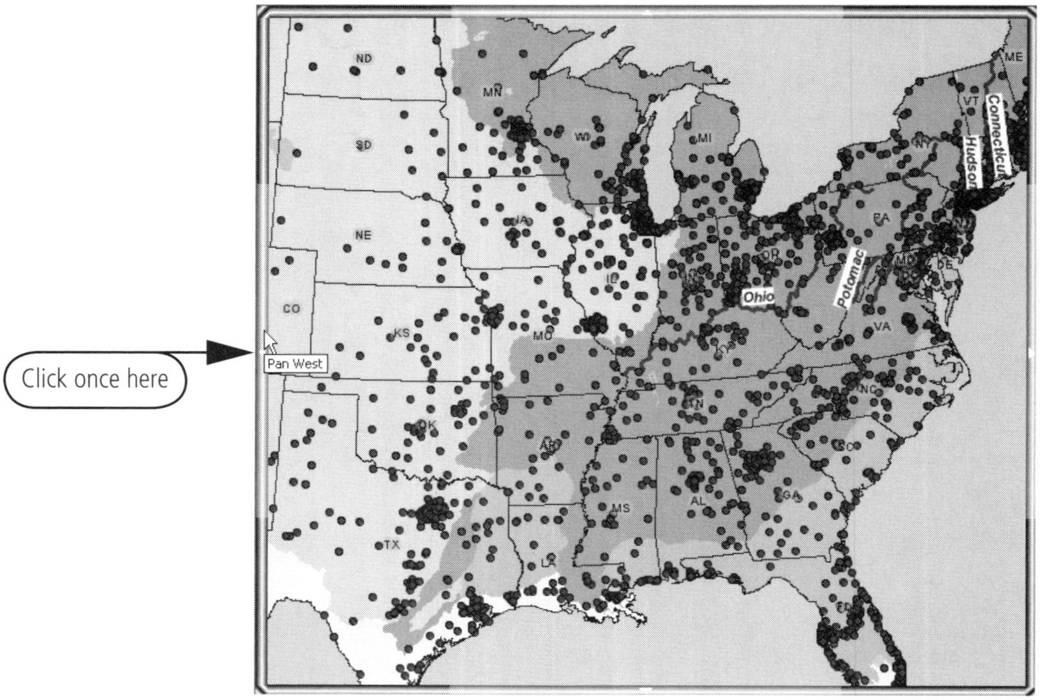

The map moves to show more of the West.

Q16 **Look at the Vegetation legend. What color represents grasslands? (Circle the correct answer.)**

a. Green

b. Pink

c. Blue

d. Orange

Step 9: The Mississippi River and the rivers that flowed into it (tributaries) offered easy routes to the central (interior) grasslands.

Once again, settlers were drawn westward in hopes of making a better living and improving their lives. West of the Mississippi, European settlers could find more ways to make a living than just farming. There were minerals like gold and silver for mining, forests for harvesting trees (in the Far West), and vast lands for raising crops and grazing cattle.

1. Turn off **NERivers** by clicking the check mark next to the layer name.

2. Click the **Add Data** button.

3. Navigate to your **OurWorld1\Module3\Data** folder.
4. Click **Rivers.shp**.

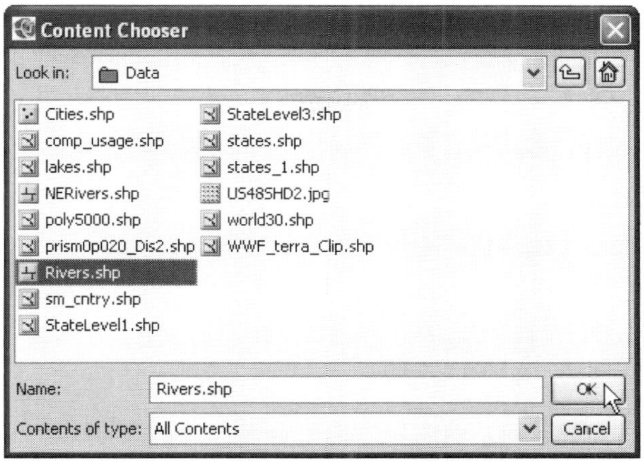

5. Click **OK**.
6. Right-click the **Rivers** layer name and choose **Properties**.
7. Change the **Color** to **Blue**.
8. Increase the **Size** to **2**.

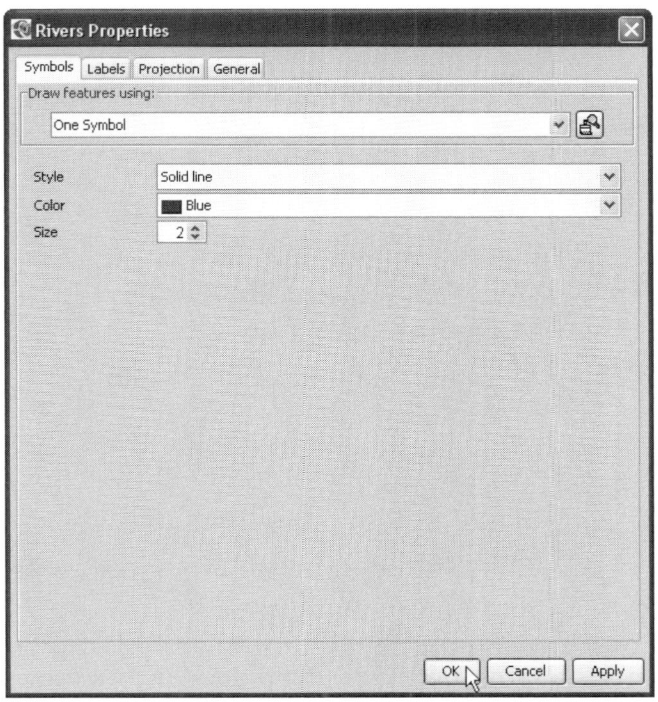

9. Click **OK**.

You see rivers of the central states. You will select the Mississippi River and its tributaries.

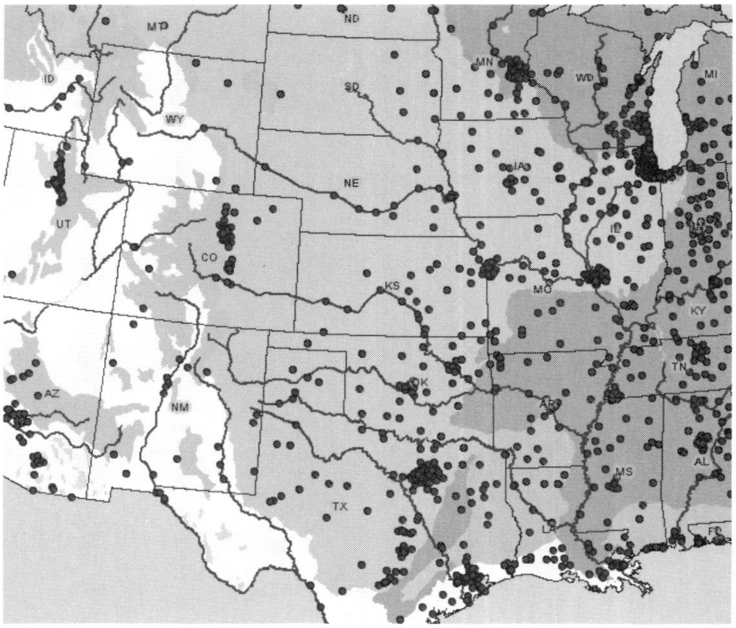

10. Click the **Rivers** layer name to highlight it.
11. Click the **Query Builder** button. The Query Builder window opens.
12. Under **Select a field**, click **SYSTEM**.
13. Click the = (equals) button.
14. Under **Values**, click **Mississippi**.

    Your query should look like this: (SYSTEM = 'Mississippi')

15. Click the **Execute** button.

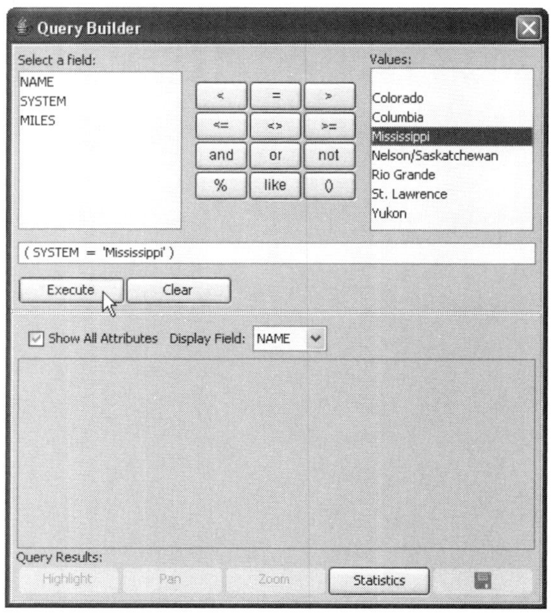

Q17 **Look at the Query Results at the bottom of the Query Builder window. How many rivers make up the Mississippi River system?** ______________________

16. Close the Query Builder window.

The Mississippi River and its tributaries are highlighted in yellow on the map.

Q18 **Look at the grasslands area on your map. As you move west across the grasslands, the pattern of cities changes. Describe these changes in your own words. Look back at Q10 and see if you can match any of the patterns.**

____

____

____

____

Step 10: Settlement was scattered from the Rocky Mountains westward.

By the 1840s, settlement had moved only a few hundred miles west of the Mississippi River.

Between 1841 and the late 1860s, westward settlement jumped across the middle of the country to Oregon and California on the Pacific Coast.

M3
L1

In January 1848, gold was discovered in California. This discovery sparked the gold rush of 1849.

- Thousands of people traveled to California using overland trails.
- Just as many people came by ship.
- Over the next few years, hundreds of thousands more people crossed the continent for gold.
- Few people got rich, and many returned home.
- Enough stayed to establish California as a state in 1850.

1. Click the small white arrow along the left border of the map to **Pan West**.

   Now you can see all the way to the West Coast.

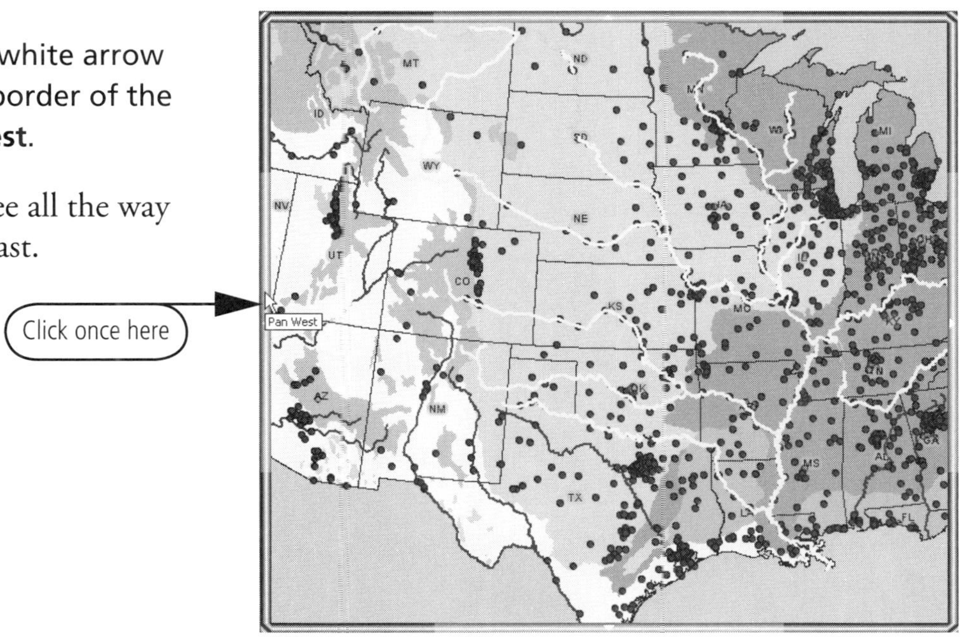

2. Click the **Pan** tool.
3. Place the cursor (which looks like a hand) over the UT label for Utah.
4. Click and *hold* down the mouse button.
5. Move the hand down until you can see all the western states.

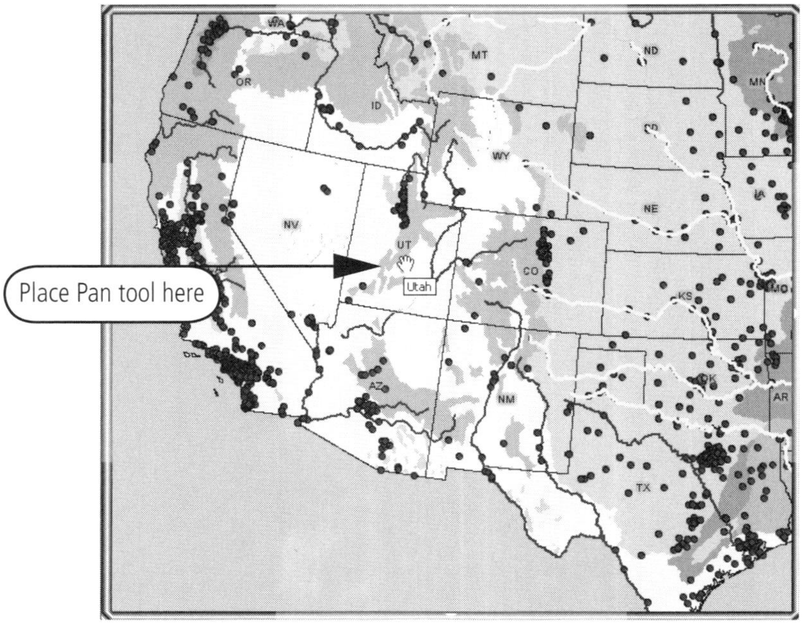

Q19 **Look at the map and look at the Vegetation legend. List three western states that are mostly covered with deserts:**

1. ______________________________

2. ______________________________

3. ______________________________

Q20 **Which pattern most closely matches the pattern of cities in these states? (Circle the correct answer.)**

a. Regular 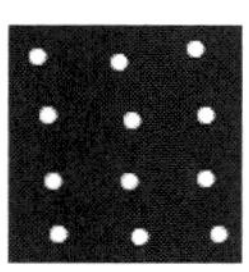

b. Clustered 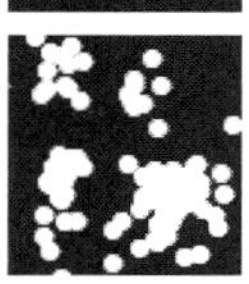

c. Random 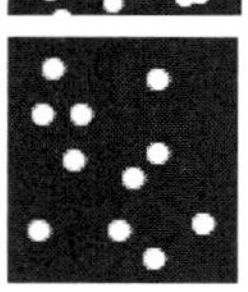

6. Turn off **Vegetation** by clicking the check mark next to the layer name.
7. Turn off **Rivers**.
8. Turn off **Cities**.
9. Turn on **Average Annual Rainfall** by clicking the box next to the layer name.

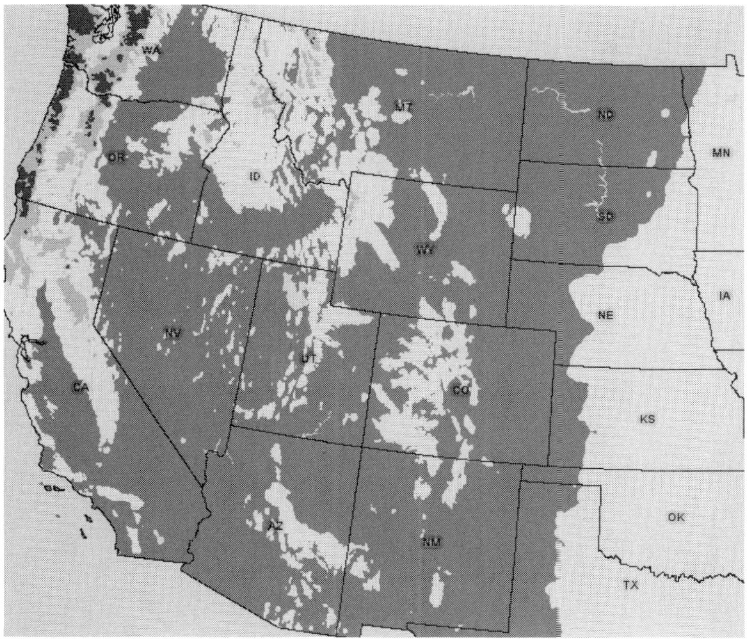

Q21 **Look at the map, and look at the Average Annual Rainfall legend. How much rain falls each year in most of the western states? (Circle the correct answer.)**

a. **Less than 22.5 inches**

b. **22.5–40 inches**

c. **40–60 inches**

So far, you have learned that some of the western states are deserts and that most of the western states have little rainfall. Let's look at elevation next.

10. Turn off **Average Annual Rainfall** by clicking the check mark next to the layer name.
11. Turn on **Elevation**.

Q22 **Look at the map, and look at the Elevation legend. What is the elevation of most of the western states? (Circle the correct answer.)**

a. **400–800 meters**

b. **800–2,000 meters**

c. **2,000 meters and greater**

12. Turn on the **Landforms.jpg** layer. After the map redraws, you see landforms (mountains, valleys, and plains) and elevation at the same time.
13. Turn on the **Cities** layer again.

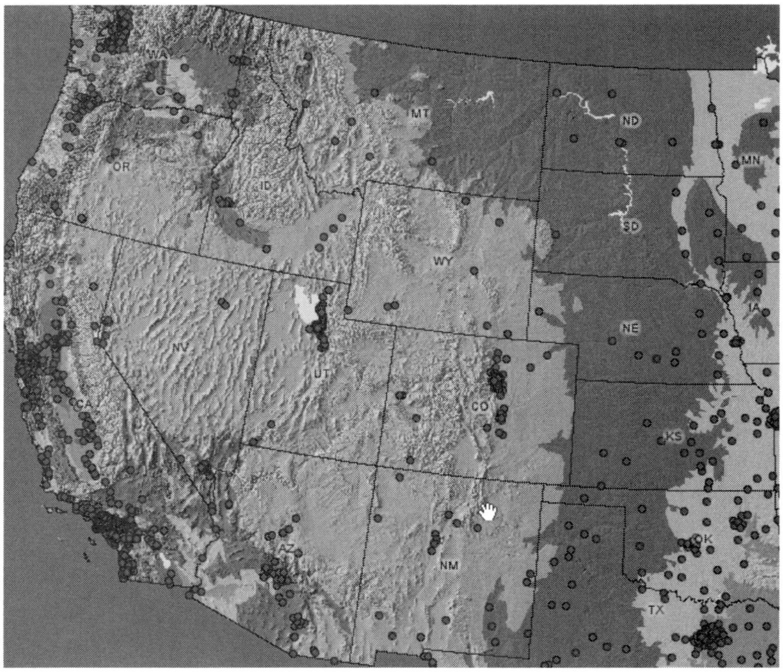

Q23 **Look at the elevation layer and the landforms layer on the map. You see mountains in high elevation areas. Which pattern most closely matches the pattern of cities in these areas? (Circle the correct answer.)**

a. **Regular** 

b. **Clustered** 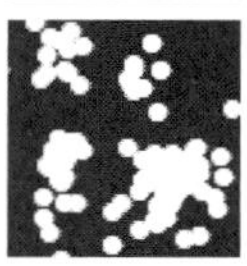

c. **Random** 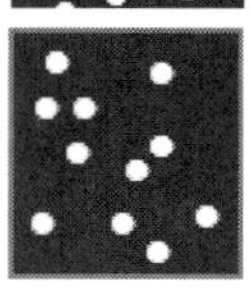

14. Click the **Lakes** layer name to highlight it. It becomes active.

 15. Click the **Zoom to Active Layer** button.

The map zooms out.

Q24 **Notice how the number and pattern of cities change as you move west across the United States. Describe these changes in your own words. Look back at Q23 and see if you can match any of the patterns.**

____

____

____

Q25 **List two reasons why settlement patterns are different in the West than in the East.**

1. ____

2. ____

M3
L1

Step 11: Save your work and exit AEJEE.

1. Ask your teacher how to save your work.
2. Click the **File** menu and choose **Exit**.

Conclusion

After the arrival of Europeans, the settlement patterns on lands that became the United States began with people moving from east to west. People moved westward as farmland increasingly became available. Then during the Industrial Revolution (1820–1870), people moved from farmlands to cities to find jobs. Once most of the jobs were in cities, people began moving from city to city. Since the 1970s, people have been moving away from cities for a variety of reasons—to places like the South, where the population had not increased for a long time.

Name: ______________________________

Date: ______________________________

Module 3, Lesson 2

Patterns of a growing population

The United States population has grown quickly during the past several hundred years. Keeping track of the nation's population dates to the country's origins. The U.S. Constitution adopted in 1787 called for a population count every 10 years, starting in 1790. This process, called the census, would keep track of the population, its activities, and its movements. More importantly, the census would ensure that each state received fair and accurate representation in the U.S. House of Representatives.

M3
L2

The 1790 Census recorded almost 4 million people. By comparison, the 2000 Census counted almost 300 million. That's more than 70 times the number of people that lived in the United States 210 years ago! It is estimated that by 2050 there will be 392 million people living in the United States! The United States now is the third most populated country in the world after China and India.

So, what kind of questions is the U.S. Census Bureau asking? The census used to ask simple questions, such as how many people lived in a home and how many cows did a family own. But the census increasingly has become a way to understand where and how government spends money in such areas as education, health care, and housing. Today, the census asks questions about many aspects of life in the United States. These questions range from wanting to know how much crime we have to how much grain, fruits, and vegetables we produce—and where we produce them.

Let's explore some of the characteristics of the more than 300 million people who live in the United States today! We will also examine how your state compares to the rest of the country. First, let's learn a little bit about where you live.

(Q1) What is the name of your state? Write it on your "Patterns of a growing population" worksheet.

A lot of decisions are made based on the information collected in the census. Let's look at some of the patterns that showed up based on how we answered questions during the 2000 U.S. Census.

Step 1: Start AEJEE.

1. Ask your teacher how to start the AEJEE software.
2. Click the **Maximize** button at the top of the window. Now the AEJEE window fills your screen.

Step 2: Open the project.

1. Click the **Open** button.
2. Navigate to your **OurWorld1\Module3** folder.
3. Choose **USCensus.axl**.
4. Click **Open**.

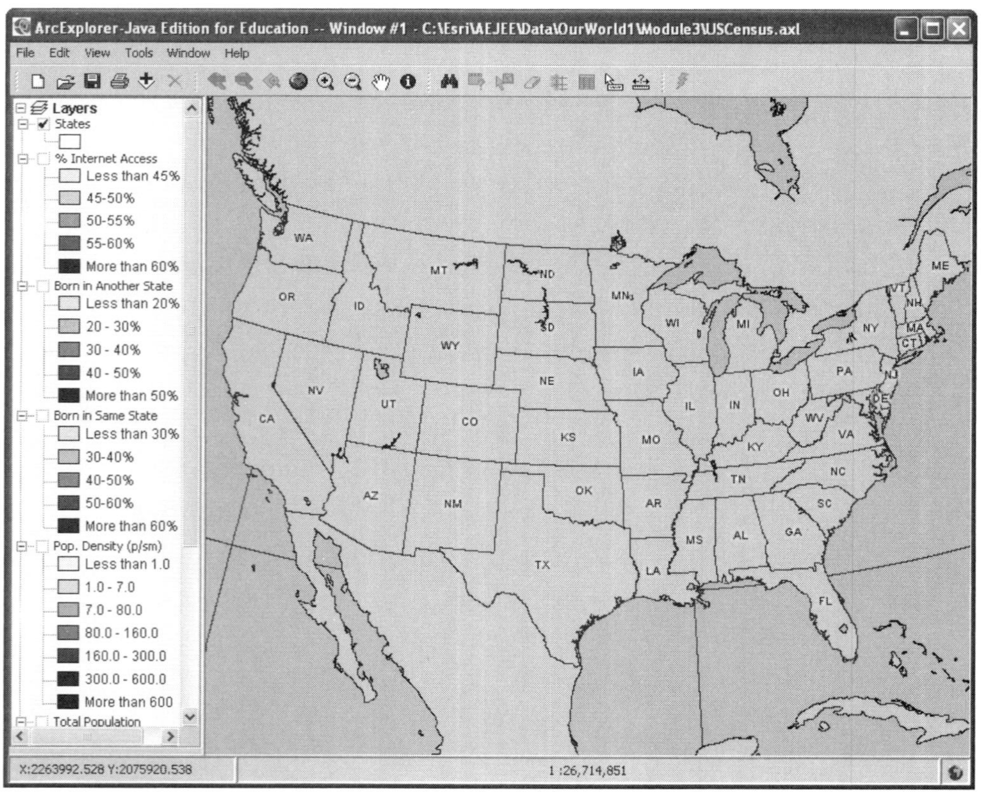

A map of the United States will be on your screen along with other background information. A list of layers is on the left side of the map. Layers are used to show geographic data on a GIS map. Each layer has a name and a legend. You can turn each layer on and off. At the top of the map, you see buttons and tools that you will use during this activity.

Step 3: Where everyone lives.

As mentioned, one of the reasons for the census is to see where everyone in the country lives. During the first census, marshals rode on horseback across the country and collected information by talking with people personally. At that time, most of the population lived on the East Coast in the original 13 states.

Things are much different today.

1. Click the name of the **States** layer to make it active. It will become highlighted.

2. Click the **MapTips** button. This will open the MapTips window.
3. Under **Layers**, click **States** (you might have to scroll down).
4. Under **Fields**, click **STATE_NAME**.
5. Click **Set MapTips**.
6. Click **OK**. Now when you hold your cursor over a state (do not click), its entire name will appear.

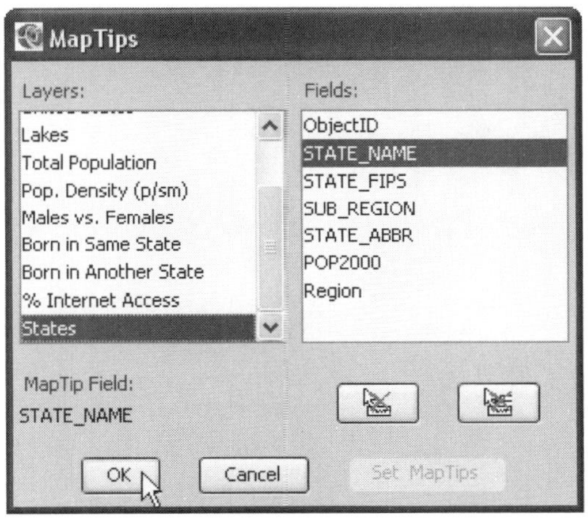

Note: If you live in Alaska or Hawaii, click the **Zoom to Active Layer** button to view your state. You can leave the map in this position for the remainder of the lesson. To go back to a view of the 48 states, click the **Zoom to Previous Extent** button.

7. Scroll down the list of layers. Turn on the **Total Population** layer by clicking the box next to its name.

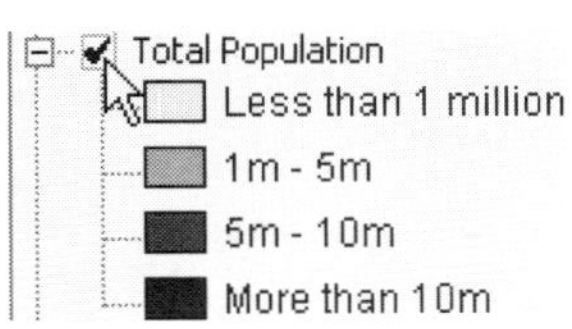

8. Click once on the **Total Population** layer name to make it active. It will become highlighted.

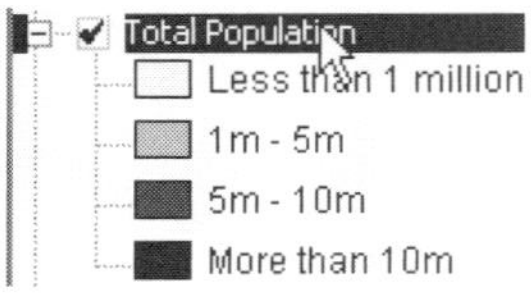

This is a *graduated color* map showing total population for the country. States with darker colors have higher populations. States with lighter colors have lower populations.

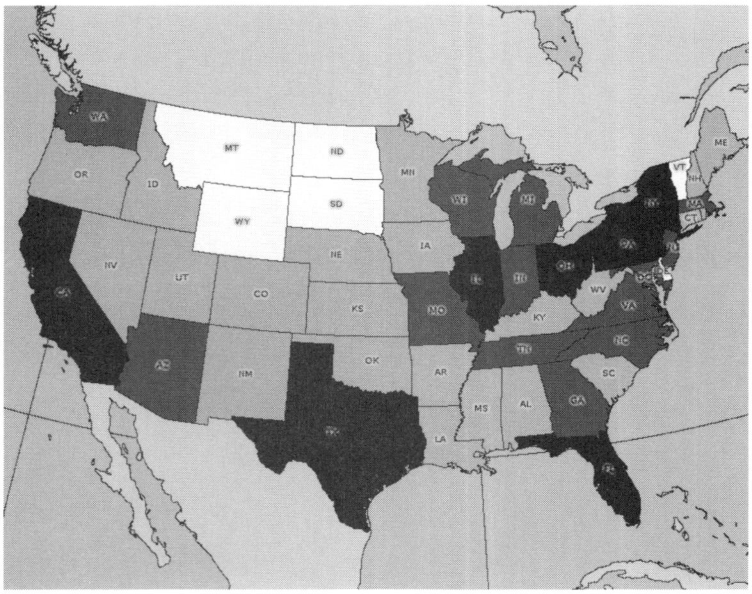

Q2 **Look at your map and find the darkest colored states. From those states, select three that each have populations of more than ten million people.**

1. ____

2. ____

3. ____

Q3 **Look at the location of all the states that fall in the highest population category. What do they have in common? (Circle the correct answer.)**

a. They are all next to one another.

b. They all start with the letter C.

c. They all have some of their border along water.

d. They are all in the middle of the country.

Q4 List three of the states that have fewer than one million people (the lightest color)?

1. ______________________________

2. ______________________________

3. ______________________________

Q5 What pattern do these states show?

a. They are mostly clustered together.

b. They are randomly spread apart.

c. They are evenly distributed across the country.

M3 L2

9. Right-click the **Total Population** layer and choose **Attribute Table**. The attribute table for this layer will appear.

10. Scroll to the right and find the **POP2000 (population in the year 2000)** field.

11. Right-click the **POP2000** field and choose **Sort Descending**. This will rearrange the records in order from the highest population to the lowest population.

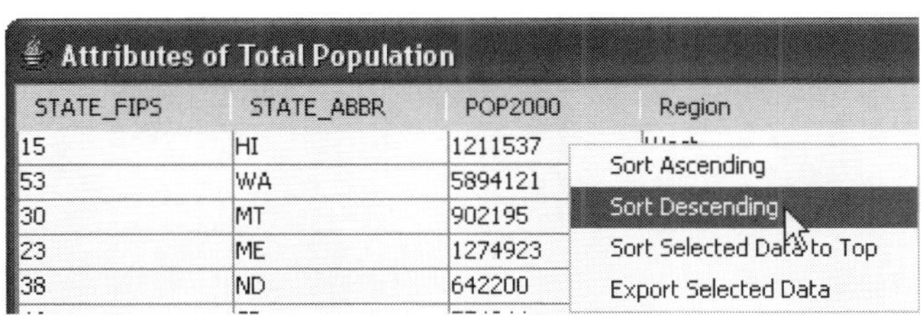

Q6 **Scroll back to the left to see the state names. Which state has the highest population (the one at the top of the list)?** ____________________

12. Right-click the **POP2000** field and choose **Sort Ascending**. This will rearrange the records in order from the lowest population to the highest population.

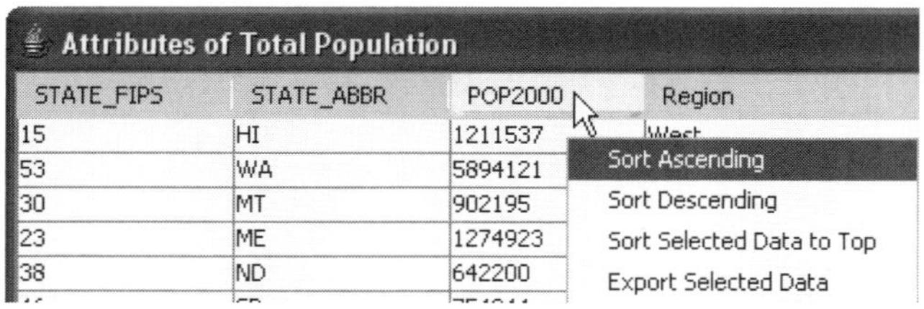

Q7 **Scroll back to see the state names. Which state has the lowest population (the one at the top of the list)?** ____________________

13. Close the **Attribute Table**.

Let's see where your state fits in.

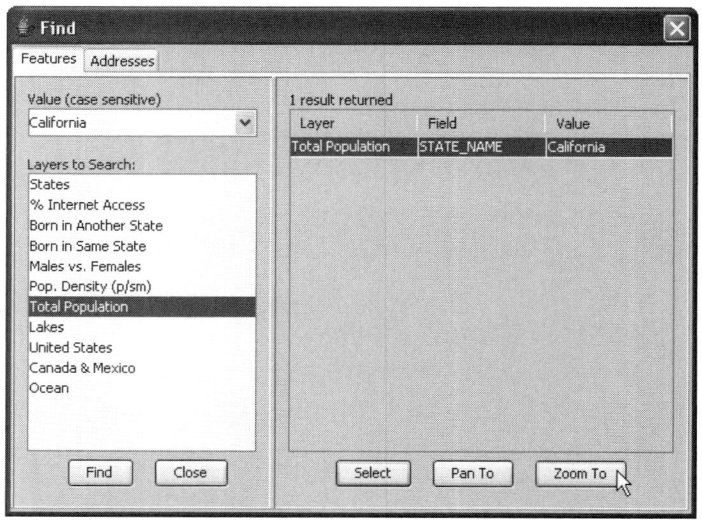

14. Click the **Find** tool.
15. Under **Value**, type the name of your state. You must type it with a capital letter at the begining and make sure you have the correct spelling.
16. Under **Layers to Search**, click **Total Population**.
17. Click **Find**. Your state will appear in the list on the right.
18. Click **Select**.
19. Click **Zoom To**. This will zoom you right into your state.

> **Note:** If your state does not appear, check your spelling and make sure the first letter is capitalized.

20. Close the **Find** window. You will be zoomed in to your state, and it is highlighted in yellow.
21. Click the **Identify** tool.

22. Click your state.

23. In the window that appears, locate the **POP2000** field.

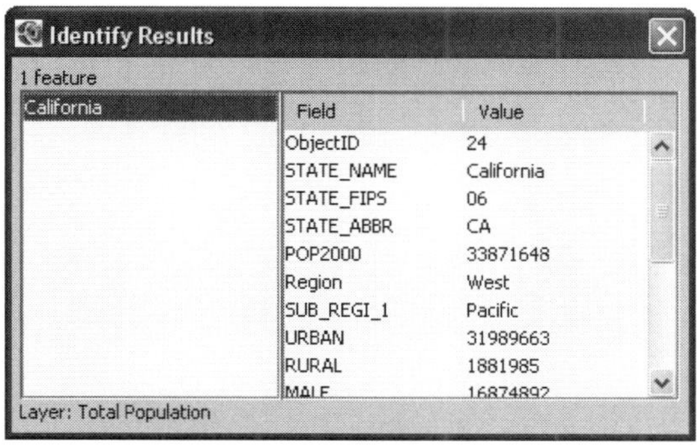

Q8 **Write the population of your state in the appropriate location on your worksheet.**

24. Close the **Identify Results** window.

25. Click the **Previous Extent** button. This will zoom you out to the continental United States.

M3 L2

Step 4: More people, less space.

In the previous step, you saw a map that showed the distribution of the total U.S. population. Another way that analysts look at population is to look at *density*.

Population density is the number of people in a particular area or unit. Usually, this is shown in terms of the number of people per square mile. In Los Angeles County, California, about 2,300 people live in each square mile. In Yellowstone County, Montana, about 49 people live in each square mile. Therefore, Los Angeles has roughly 47 times more people per square mile!

1. Turn off the **Total Population** layer by clicking the check mark next to its name.

2. Turn on the **Pop. Density (p/sm)** layer by clicking the box next to its name.

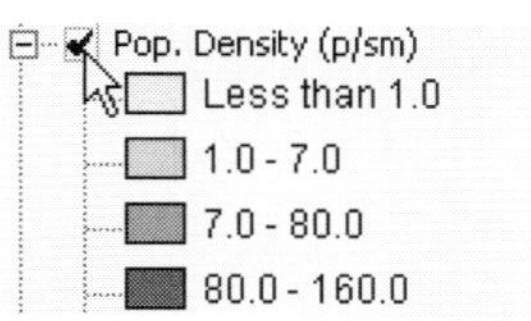

Let's find the states with the highest and lowest population densities.

3. Click the **Pop. Density (p/sm)** layer name to make it active. It will become highlighted.

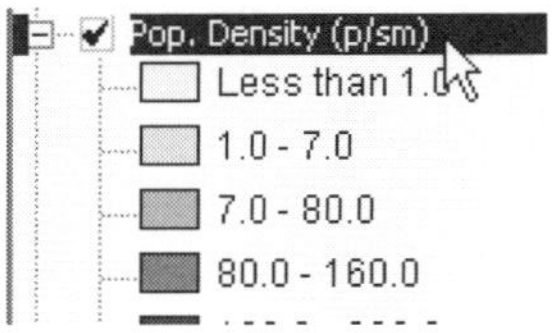

4. Click the **Zoom to Active Layer** button. This will zoom the map out to the entire country, including Alaska and Hawaii. (Hint: If you were already zoomed out, then you do not need to do this again.)
5. Right-click the **Pop. Density (p/sm)** layer name.
6. Choose **Attribute Table**.

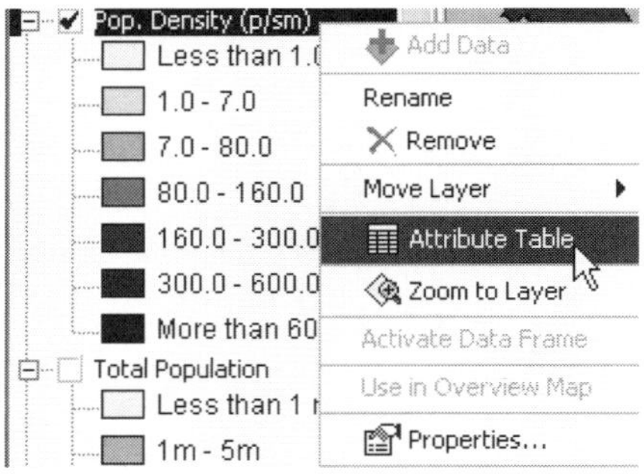

7. Scroll to the right until you find the **Pop_Dens** (population density) field.
8. Right-click the **Pop_Dens** field and choose **Sort Ascending**. This will rearrange the records in the table from lowest to highest population density.

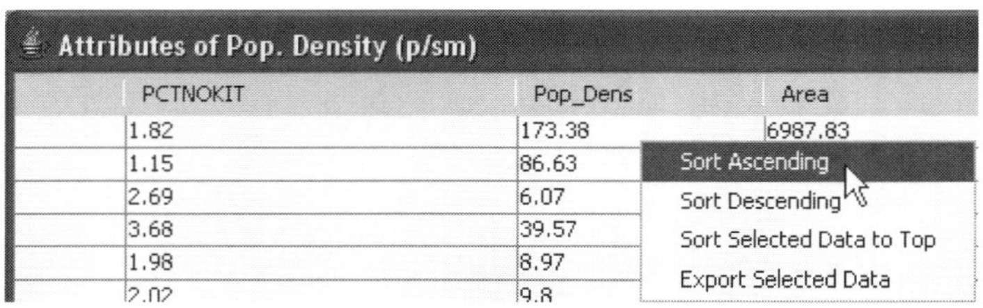

9. Click the first record in the list. It will be highlighted in blue.

Q9 **What is the name of the state that has the lowest population density?**

State: ______________________________

10. Close the **Attribute Table**.

Q10 **Can you think of two reasons why this state has the lowest population density?**

1. ______________________________

2. ______________________________

11. Click the **Previous Extent** button. This will bring you back to a map of the continental United States. (If you live in Alaska or Hawaii, do not click **Previous Extent**.)
12. Right-click **Pop. Density (p/sm)** again.
13. Choose **Attribute Table**.
14. Scroll to the right and locate the **Pop_Dens** field.
15. Right-click the **Pop_Dens** field and choose **Sort Descending**. This will rearrange the list from highest to lowest population density.

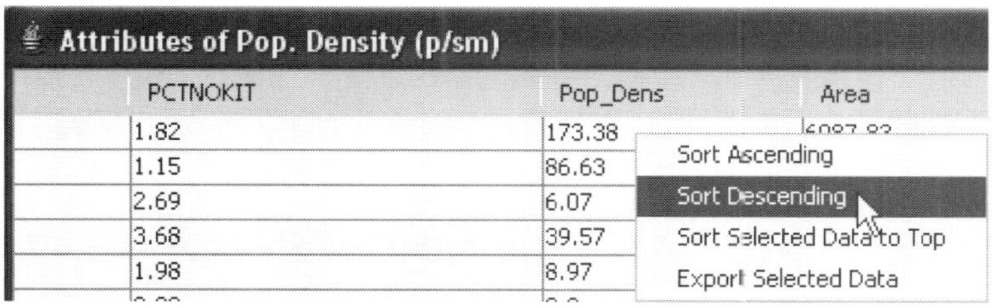

M3
L2

16. Click the first record. It will be highlighted in blue in the table and in yellow on the map.

Q11 **Scroll back to see the names of the states. Which state has the highest population density?**
State: ______________________________

Population density: ______________________________ **people per square mile**

Q12 **Can you think of two reasons why this state has the highest population density?**
Answers may include the following:

1. ______________________________

2. ______________________________

17. Close the **Attribute Table**.

18. Click the **Clear All Selections** button.

Q13 **Look at your map and notice how population density is spread across the states. Describe any patterns you see. Use words such as "located together" (or clustered), "without any pattern" (or random), and "looking the same as the others" (or uniform).**

Let's see where your state fits in.

19. Click the **Identify** tool.
20. Click your state. (Hint: If it's hard to find your state, hold your cursor over the state until the name pops up in the MapTip. Then click your state.)
21. In the window that appears, scroll down and locate the **Pop_Dens** field.

Q14 **What was the population density (Pop_Dens) of your state at the 2000 Census? Write it on your worksheet.**

22. Close the **Identify Results** window.

Step 5: Moving around the country.

People who live in the United States are on the move. Did you know that Americans on average move eleven times in their lives? While some people never move, about one in five people move each year—that's 20 percent of the population. Some move for family reasons such as getting married. Others move to start new jobs, go away to college, or buy a different house.

It's not easy keeping track of population movement. The U.S. Census asks people their current address but also asks if they have moved in the last five years.

1. Click the **Previous Extent** button. You will go back to a view of the continental United States. (Hint: If you are already at this view, then you do not need to click the button.)
2. Turn off the **Pop. Density (p/sm)** layer.
3. Turn on the **Born in Same State** layer. This layer shows the percent of people that were born in the same state they currently live in.

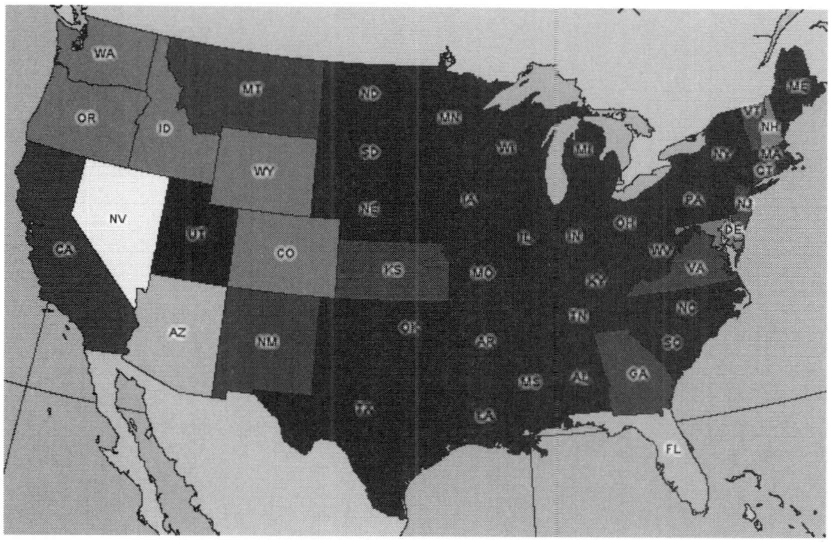

Let's look at the states in the highest category.

4. Click the **Born in Same State** layer name to highlight it. It becomes active.

5. Click the **Query Builder** button.
6. Under **Select a field,** choose **PCTINSTATE**.
7. Click the >= (greater than or equal to) button.
8. Put your cursor next to the >= sign and type **70**.
9. Your query should look like this: **(PCTINSTATE >= 70)**
10. Make sure the parentheses are there!
11. Click **Execute**. The states that have more than 70 percent of people who are born in the state will appear in the list at the bottom of the window. They will also appear in yellow on your map.

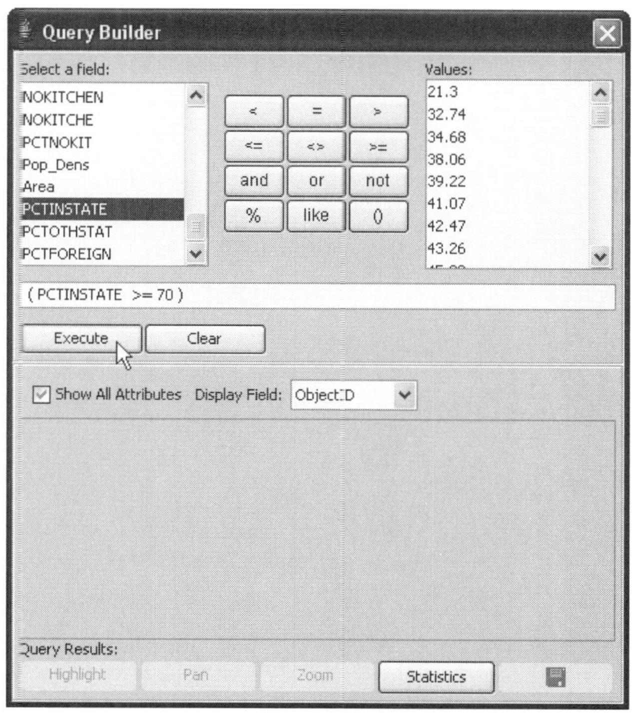

Q15 **Select any five states from those that become selected and write their names here:**

1. ____________________

2. ____________________

3. ____________________

4. ____________________

5. ____________________

12. Close the **Query Builder**. The states will still be highlighted.

Q16 **In which part of the country are most of these states located?**

a. East

b. West

Q17 **Discuss with your classmates why you think people want to remain living in these states.**

Q18 **List two reasons why people would not move away from the state they were born in.**

1. ____________________

2. ____________________

Q19 **Ask your classmates if they were born in the state they live in. How many said yes? How many said no?**

Yes____

No____

13. Click the **Clear All Selections** button.

Let's see where your state fits in.

14. Click the **Identify** tool.
15. Find your state on the map.
16. Click your state. The **Identify Results** window appears.
17. Scroll down until you see the **PCTINSTATE** field.

Q20 **What is the percent of people living in the state that were born there? Write this on your worksheet.**

M3
L2

18. Close the **Identify Results** window.
19. Turn off the **Born in Same State** layer.
20. Turn on the **Born in Another State** layer.

This map shows the percent of people in each state who were born in a different state.

Q21 **Did you notice a significant change in the map? (Hint: Turn off and on the Born in Another State layer to take another look.)**

a. Yes

b. No

Q22 **Describe the change in pattern that you noticed.**

1. ____

2. ____

21. Turn off the **Born in Same State** layer.

(Q23) Look at the map. Which states have the highest number of people who were born in another state (the darkest color)? Write two of them here.

1. ____

2. ____

(Q24) What do you think makes these states so popular?

____

____

____

22. In the list of layers, find the **Born in Another State** layer.
23. Click the colored box that represents the *lowest* number of people who were born in another state (Less than 20%). All the states that fall in that category will be highlighted in green on the map.

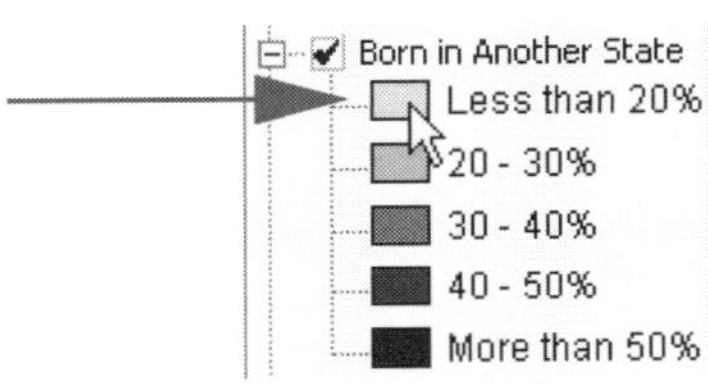

(Q25) Look at the map. Which states have the lowest number of people who were born in another state (the lightest color)? List three of them here:

1. ____

2. ____

3. ____

Q26 **Why do you think so few people from other states move there?**

24. Click the **Born in Another State** layer name to highlight it. It becomes active.
25. Click the **Clear All Selection** button.

Let's see where your state fits in.

26. Click the **Identify** tool.
27. Find your state on the map.
28. Click your state. The **Identify Results** window appears.
29. Scroll down until you see the **PCTOTHSTAT** field.

M3 L2

Q27 **What is the percent of people living in your state who were born in another state? Write this on your worksheet.**

30. Close the **Identify Results** window.

Step 6: Surfing the Internet.

Each year, the U.S. Census performs a survey called the American Community Survey. The survey collects interesting facts on the U.S. population and publishes them on the Web.

One of the pieces of data the survey examines is people's access to the Internet. The U.S. Census collects this information for each household in the United States.

1. Turn off the **Born in Another State** layer.
2. Turn on the **%Internet Access** layer by clicking the box next to the layer name.

Q28 **Look at your map. List three of the states where more than 60 percent of households have access to the Internet? (These will be the states in the darkest colors.)**

1. ______________________________

2. ______________________________

3. ______________________________

Q29 **Look at your map. What pattern do you notice among the states where more than 60 percent of the population has Internet access? (Circle all that apply.)**

a. **There is a random pattern—these states are spread all over the country.**

b. **There is a cluster pattern—the states are grouped together in certain areas.**

c. **There is a uniform pattern—the states are spaced evenly apart.**

3. Click the **%Internet Access** layer name to highlight it. It becomes active.
4. Right-click the **%Internet Access** layer and choose **Attribute Table**.
5. Right-click the **INTACC_PCT** field. This is the percent of households with access to the Internet.
6. Choose **Sort Ascending**. This puts the states in order from lowest to highest percentage.
7. Hold down your **CTRL** key on the keyboard, and click the first three states in the table. They become highlighted in blue. They will also become highlighted on your map.

Attributes of % Internet Access

COMP_TOT	INTACC_TOT	INTACC_PCT	COMACC_PC
537	435	39.55	48.82
544	462	42.31	49.82
935	785	44.1	52.53
390	323	44.86	54.17
996	846	45.8	53.93
899	749	45.95	55.15
409	354	47.39	54.75
787	687	48.31	55.34
1357	1171	49.02	56.8
951	813	49.91	58.38

Selected:3

> **Note:** If you cannot see the map underneath, move the attribute table window out of the way.

Q30 **Which three states have the lowest percent of households with access to the Internet? You might have to scroll back to the left to see the names of the states.**

1. ____________________

2. ____________________

3. ____________________

Q31 **In which region of the country are these states located?**

a. **North**

b. **South**

c. **East**

d. **West**

M3
L2

8. Scroll to the bottom of the table.
9. Click the **last state** (row) in the table (don't use the CTRL key).
10. Hold down the **CTRL** key on your keyboard and click the **two states** above it. The last three states should be highlighted in blue. They will also be highlighted on your map.

Q32 **Which three states have the highest percent of households with access to the Internet?**

1. ____________________

2. ____________________

3. ____________________

11. Close the **Attribute Table**.
12. Click **Clear All Selections**.

Q33 **Overall, which part of the country appears to have the most access to the Internet?**

a. North

b. South

c. East

d. West

Once again, let's look at where your state fits in.

13. Click the **Identify** tool.
14. Find your state on the map.
15. Click your state. The **Identify Results** window appears.
16. Scroll down until you see the **INTACC_PCT** field.

Q34 **What is the percent of Internet access for households in your state? Write this number on your worksheet.**

17. Close the **Identify Results** window.

Step 7: Save your work and exit AEJEE.

1. Ask your teacher how to save your work.
2. Click the **File menu** and choose **Exit**.

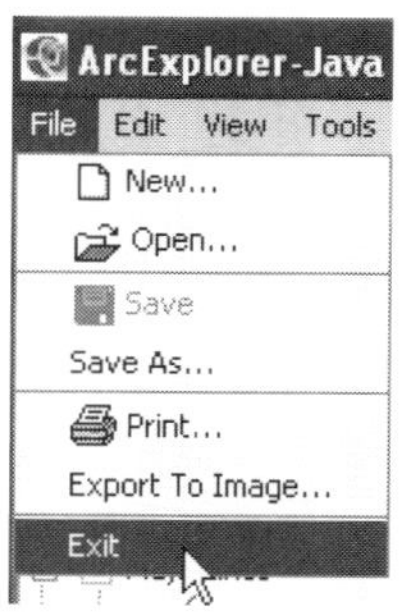

Conclusion

The U.S. Census collects a lot of information about the population. The 2010 Census will have some new collection methods, including handheld computers for enumerators (the people who go door-to-door to collect census data). Questions change with each census as the information required by the U.S. government changes. The most important thing that you can do to help the U.S. Census is to make sure that you get counted!

Using GIS to create maps of the census information can show people how things are spread out across the country.

Name: ______________________________

Date: ______________________________

Module 3, Lesson 2

Patterns of a growing population

Worksheet

Name of your state: ____

Topic	Your state
Population in 2000	
Population density	
Born in state	
Born in another state	
Internet access	

M3
L2

Once you have completed this table, write a brief paragraph that describes the population characteristics of your state. Use the information you have obtained during this activity.

____

____

____

____

____

Name: ______________________

Date: ______________________

Module 4, Lesson 1

Finding Tornado Alley

Tornado Alley is an area of the United States that has more tornadoes than any other place on earth. Many people argue about which states are in Tornado Alley. In this lesson, you will learn how to find this area for yourself! By the time you finish this lesson, you will be able to list the states that have the most frequent tornadoes, the strongest tornadoes, and the greatest concentration of tornadoes. From your list, you will be able to identify states that are in Tornado Alley.

Tornadoes are associated with certain weather patterns, and these patterns change with the seasons. In this lesson, you will learn which regions of the United States have tornadoes at different times of the year—winter, spring, summer, and fall.

Let's explore Tornado Alley.

Step 1: Start AEJEE.

1. Ask your teacher how to start the AEJEE software.
2. Click the **Maximize** button at the top of the AEJEE window. Now the AEJEE window fills your screen.

Step 2: Open the project.

1. Click the **Open** button.
2. Navigate to your **OurWorld1\Module4** folder.
3. Choose **USTornadoes.axl**.
4. Click **Open**.

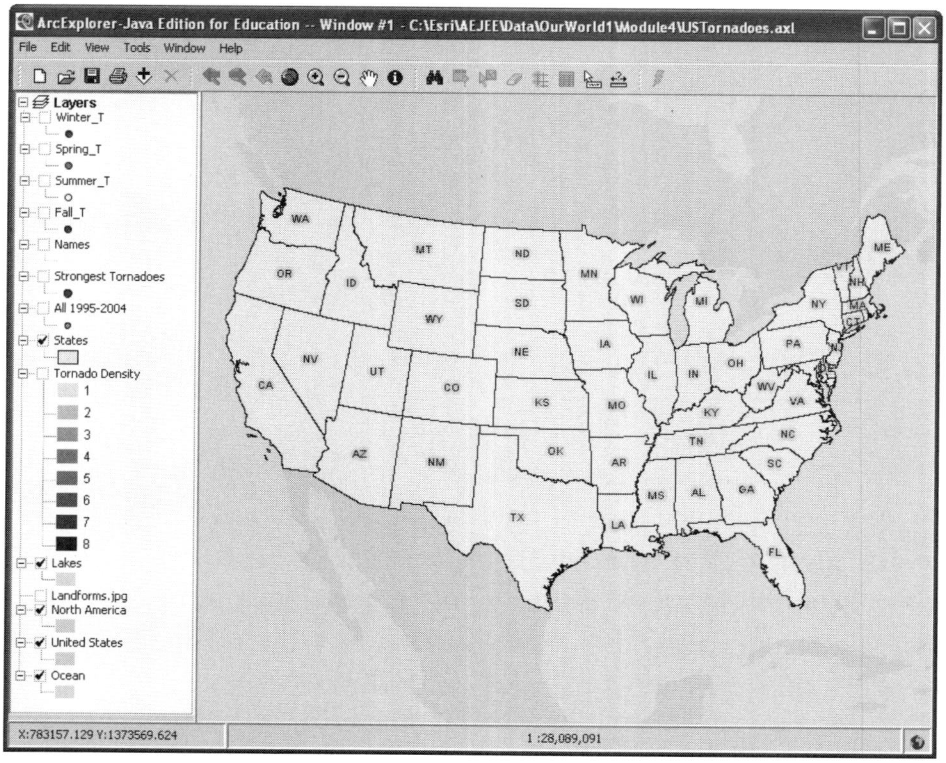

A map of the 48 states labeled with their abbreviations appears on your screen. On the left side of the map, you see a list of layers. Layers are used to show geographic data on a GIS map. Each layer has a name and a legend. You can turn layers on and off. At the top of the map, you see menus and buttons that you will use during this activity.

Step 3: Tornadoes of the United States.

1. Turn on the **All 1995-2004** layer by clicking the box next to the layer name. Each dot represents a tornado that occurred from 1995–2004 (a period of 10 years).

Q1 **Look at your map. In what part of the country do most tornadoes occur? (Circle the correct answer.)**

a. Eastern states

b. Western states

c. Central and eastern states

Q2 **Do you see any states that don't have tornadoes? (Circle the correct answer.)**

a. Yes

b. No

2. Click the **States** layer name to highlight it. It becomes active.

3. Click the **Zoom to Active Layer** button to zoom out to all the states. Now you can see Alaska and Hawaii too.

Q3 **Do you see any states that don't have tornadoes? (Circle the correct answer.)**

a. Yes

b. No

4. Click the **Previous Extent** button to zoom back in to the 48 states.

Tornadoes are more common in some parts of the country than others. Let's look at a different view of the United States.

5. Turn off the **States** layer by clicking the check mark next to the layer name.

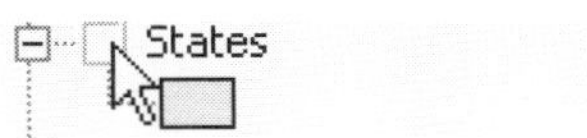

6. Turn off the **All 1995-2004** layer.

7. Turn on **Landforms.jpg** by clicking the box next to the layer name. Be patient while this layer draws.

You see an image of the United States showing high areas (mountains) and low areas (plains and valleys).

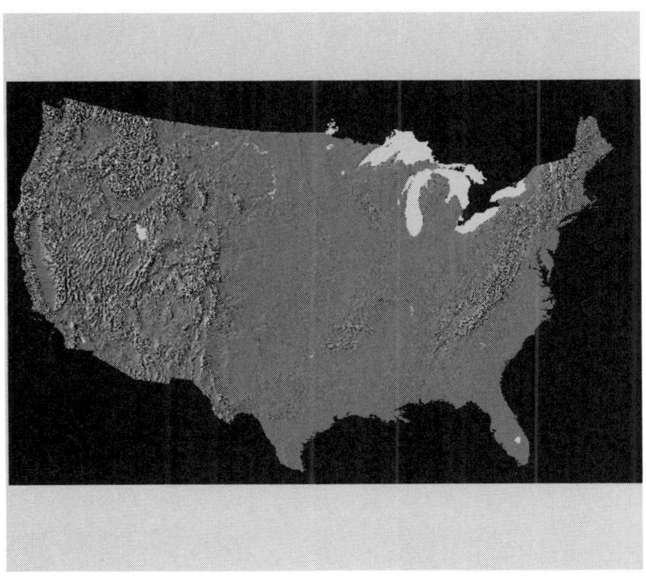

8. Turn on the **Names** layer by clicking the box next to the layer name.

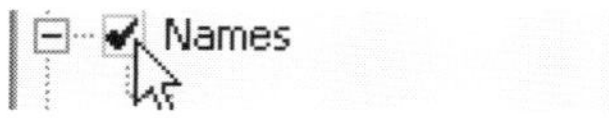

Now you see the names of some of the landforms, water bodies, and neighboring countries of the United States.

9. Turn on the **All 1995-2004** layer again.

Q4 **Look at your map. In what part of the country do most tornadoes occur? (Circle the correct answer.)**

a. Rocky Mountains

b. Great Plains

c. Appalachian Mountains

The Great Plains have lots of thunderstorms, creating the kind of weather conditions that can cause tornadoes. Thunderstorms occur when warm, moist air meets cool, dry air.

10. Turn off the **Landforms.jpg** layer.

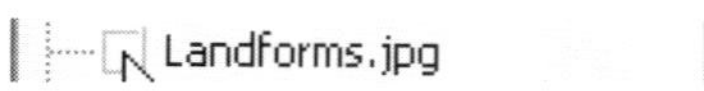

11. Turn on the **States** layer.

Warm, moist air that feeds these thunderstorms comes from the body of water south of the Great Plains.

Q5 **Look at your map. What is the name of this body of water?**____________________

Cool, dry air comes from the land to the north of the Great Plains.

Q6 **Look at your map. What is the name of this country?**________________________

12. Turn off the **Names** layer.

13. Turn off the **All 1995-2004** layer.

Step 4: States where tornadoes occur most often.

While Tornado Alley is generally in the middle of the country, many people argue about the specific states that belong in this area. There are three different ways, or methods, to decide which states are in Tornado Alley:

1. States where tornadoes occur most often
2. States where the most dangerous tornadoes occur
3. States where the greatest concentration of tornadoes occurs

You will use GIS to analyze these three methods and decide on some of the states that are in Tornado Alley. You will start by finding out where tornadoes occur most often.

1. Click the **Add Data** button.
2. Navigate to your **OurWorld1\Module4\Data** folder.
3. Click **State_tornadoes.shp**.

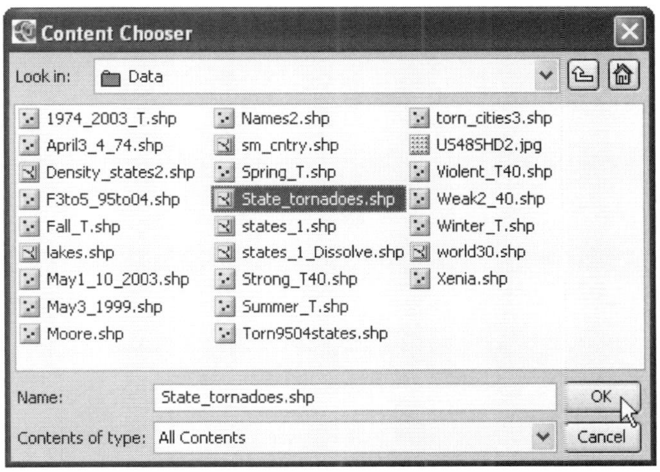

4. Click **OK**. The **State_tornadoes** layer is added to the map and draws in a random color.

   The State_tornadoes layer contains information about the tornadoes for each state. You will change its symbols.

5. Click the **State_tornadoes** layer name to highlight it. It becomes active.
6. Right-click the **State_tornadoes** layer name and choose **Properties**. This brings up the **Properties** window.

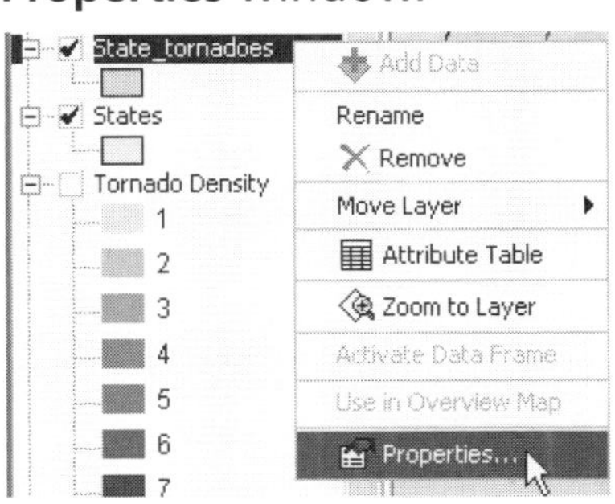

M4
L1

7. Under **Draw features using**, choose **Graduated Symbols**.

8. For **Field**, choose **AVERAGE_T**. This field contains the average number of tornadoes per year in each state.

9. Click **OK**.

You see a map showing the average number of tornadoes per year in each state (from 1995 to 2004). The red color means there were a lot of tornadoes; the yellow means there were relatively few tornadoes.

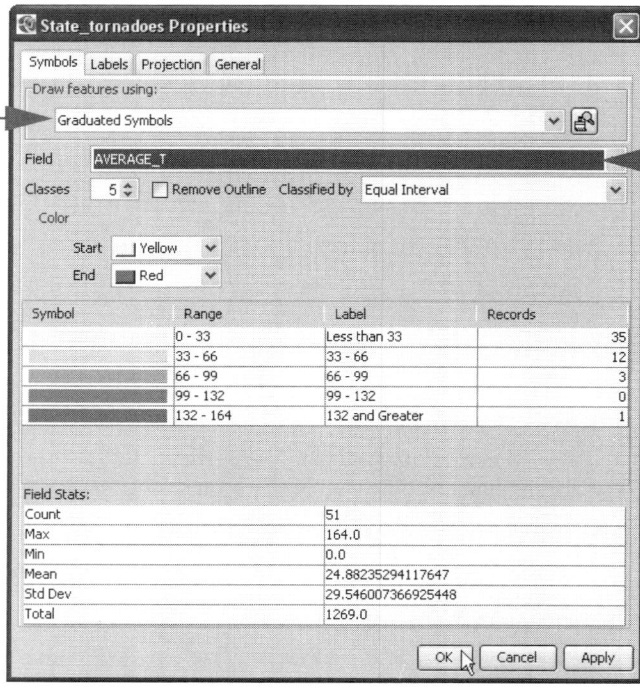

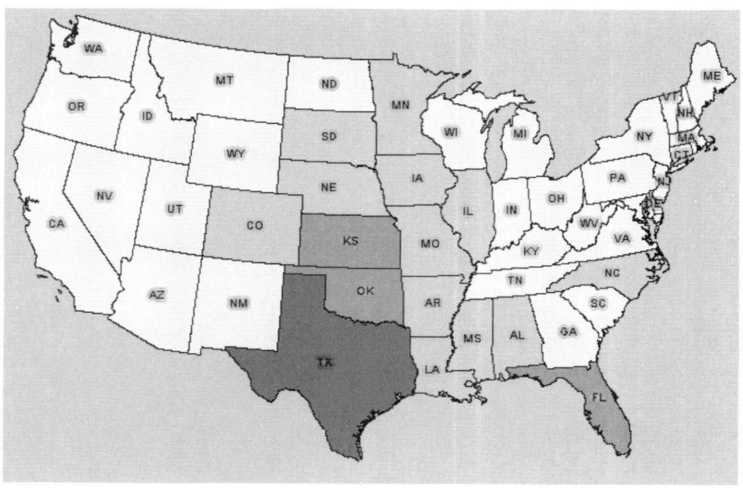

What does "average number of tornadoes per year" mean?

Suppose a state had 200 tornadoes from 1995 to 2004 (a period of 10 years). If you divide the total number of tornadoes (200) by the number of years (10), you will get the average number of tornadoes per year:

$200 \div 10 = 20$ tornadoes per year

Q7 **Look at your map and at the legend. In which state did tornadoes occur most often?** ____________________

10. Right-click **State_tornadoes** and choose **Attribute Table**. This brings up the Attribute Table.

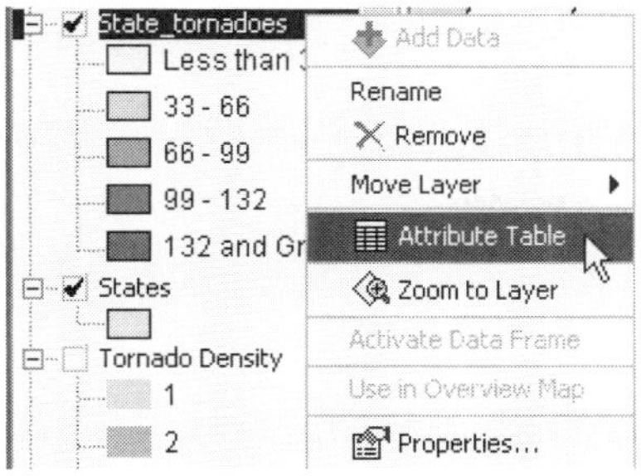

11. Scroll to the right in the table until you see the **AVERAGE_T** field.
12. Right-click the **AVERAGE_T** field name and choose **Sort Descending**. The states are listed in order from the highest number to the lowest number of tornadoes.

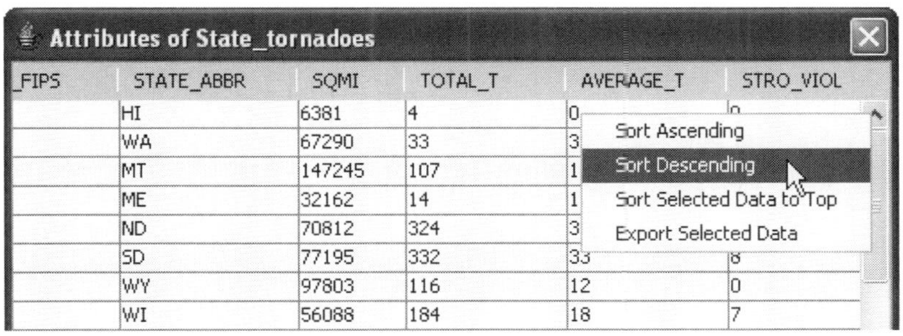

FIPS	STATE_ABBR	SQMI	TOTAL_T	AVERAGE_T	STRO_VIOL
	HI	6381	4	0	0
	WA	67290	33	3	
	MT	147245	107	1	
	ME	32162	14	1	
	ND	70812	324	3	
	SD	77195	332	33	8
	WY	97803	116	12	0
	WI	56088	184	18	7

Let's look at the top six states.

13. Hold down your **CTRL** key on the keyboard, and slowly click each of the first **six** records in the table (count them). They become highlighted in blue. They also become highlighted on your map.

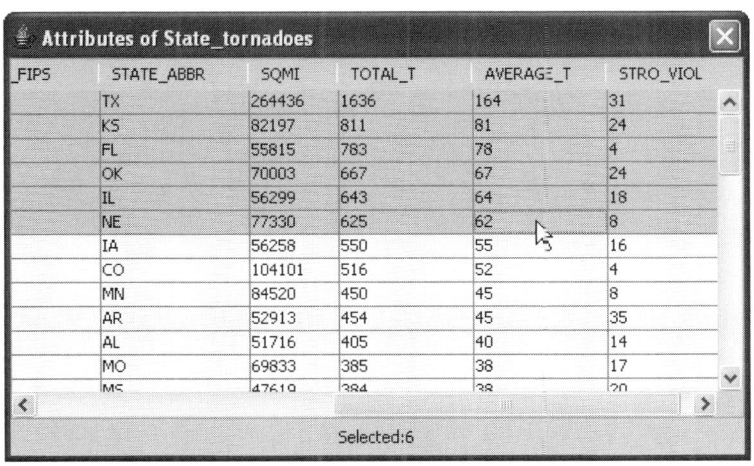

FIPS	STATE_ABBR	SQMI	TOTAL_T	AVERAGE_T	STRO_VIOL
	TX	264436	1636	164	31
	KS	82197	811	81	24
	FL	55815	783	78	4
	OK	70003	667	67	24
	IL	56299	643	64	18
	NE	77330	625	62	8
	IA	56258	550	55	16
	CO	104101	516	52	4
	MN	84520	450	45	8
	AR	52913	454	45	35
	AL	51716	405	40	14
	MO	69833	385	38	17
	MS	47619	384	38	20

Note: If you make a mistake, click the first record without holding down the CTRL key. Then hold down the CTRL key and click the next five records.

14. Let go of the **CTRL** key.
15. Scroll to the left in the table until you see the **STATE_NAME** field. You see the names of six states highlighted in blue.
16. Take out the "Tornado Alley" worksheet that your teacher gave you.
17. Find Column 1: Highest average number of tornadoes.

Q8 **Write the names of these six states in Column 1 of your worksheet. (Some state names are filled in for you.)**

18. Close the **Attribute Table**.
19. Click the **Clear All Selections** button.

Step 5: States where the strongest tornadoes happen.

Another way to decide which states are in Tornado Alley is to see where the strongest tornadoes happen. The strength of a tornado is measured by the speed of its winds and by how much damage it causes. You will look at which states had the strongest tornadoes between 1995 and 2004.

1. Turn off the **State_tornadoes** layer. The layer name should still be highlighted.

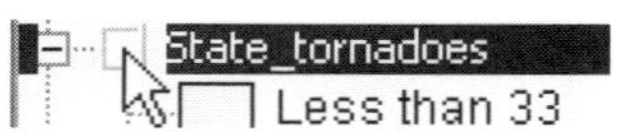

2. Turn on the **Strongest Tornadoes** layer.

Each purple dot represents a single tornado. These tornadoes had winds that were stronger than 150 miles per hour and caused a lot of damage. Only a small portion of all tornadoes are really strong.

Q9 **Look at the map. List two states that you think have a high number of strong tornadoes.**

1. ______________________ **2.** ______________________

Note: If you aren't sure of the name of a state, click the Identify tool, then click the state. Close the Identify Results window when you are finished.

Now you will find out if you were right.

3. Turn off **Strongest Tornadoes**.

4. Turn on **State_tornadoes**. The layer name should still be highlighted.

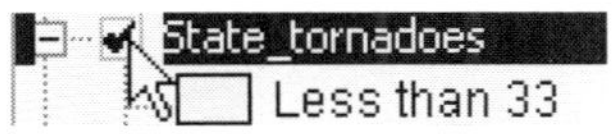

5. Right-click **State_tornadoes** and choose **Properties**. This brings up the **Properties** window.

6. Change the **Field** to **STRO_VIOL**. This field contains the strongest tornadoes from 1995 to 2004.

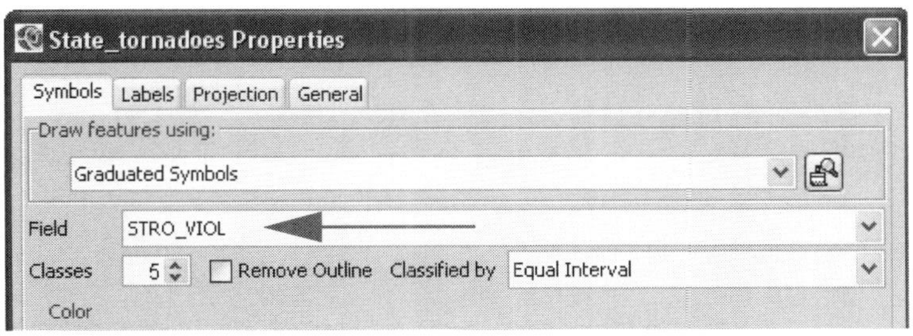

7. Click **OK**.

The red color means more strong and violent tornadoes, the yellow color means less.

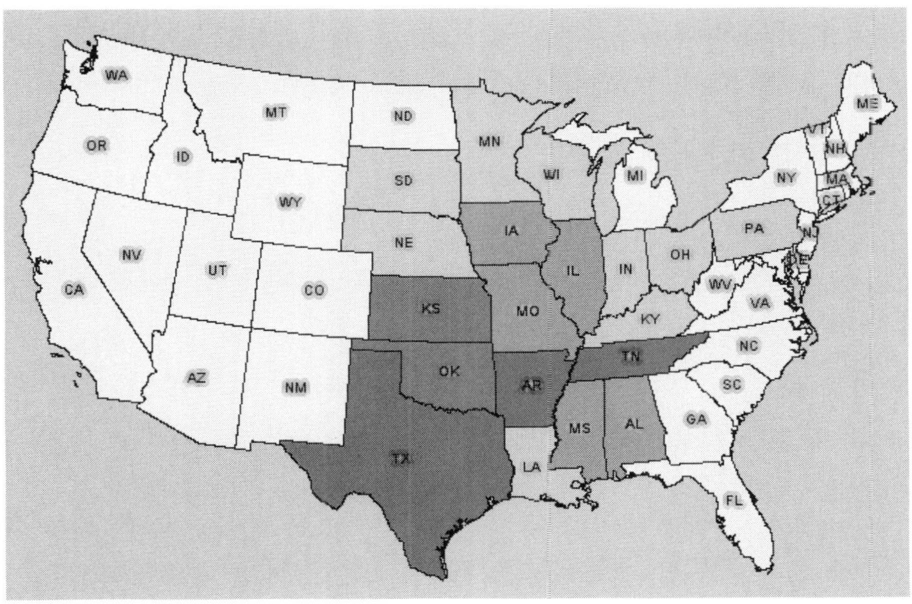

8. Click the **Identify** tool.

9. Click one of the **dark red states**. The **Identify Results** window opens. On the left side of the window you see the name of the state. On the right side you see field names and values.

M4
L1

10. Look at the value for the **STRO_VIOL** field. It tells you the number of strong and violent tornadoes that occurred in that state from 1995 to 2004.

Q10 **What is the name of this state?** ______________________________

Q11 **How many strong and violent tornadoes did it have from 1995 to 2004?**

____

11. Close the **Identify Results** window.
12. Click another **dark red state** on the map.

Q12 **What is the name of this state?** ______________________________.

Q13 **How many strong and violent tornadoes did it have from 1995 to 2004?**

____

13. Close the **Identify Results** window.
14. Right-click the **State_tornadoes** layer name and choose **Attribute Table.**
15. Scroll to the right side of the table.
16. Right-click the **STRO_VIOL** field name and choose **Sort Descending**.

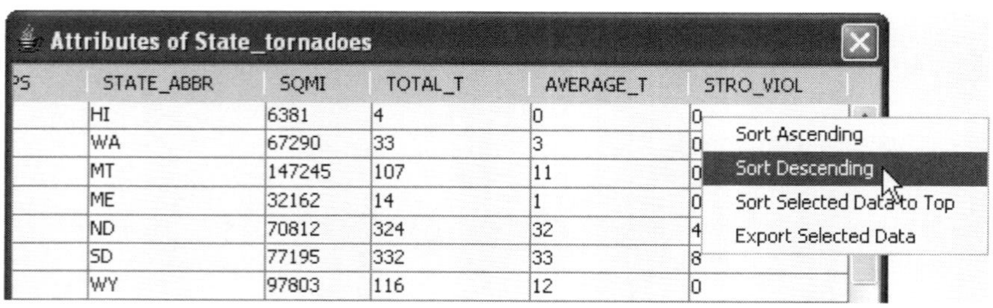

The states are listed in order from the highest number of strong and violent tornadoes to the lowest number.

Let's look at the top six states.

17. Hold down your **CTRL** key on your keyboard and click the first **six** records in the table (count them). They become highlighted in blue. They also become highlighted on your map.

> **Note:** If you make a mistake, click the first record without holding down the CTRL key. Then hold down the CTRL key and click the next five records.

18. Let go of the **CTRL** key.

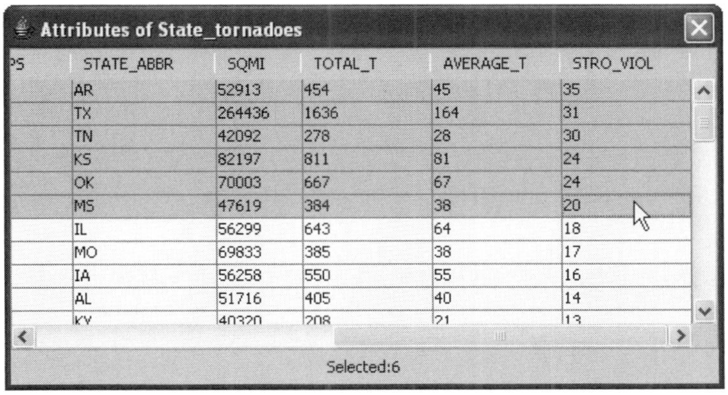

Attributes of State_tornadoes

PS	STATE_ABBR	SQMI	TOTAL_T	AVERAGE_T	STRO_VIOL
	AR	52913	454	45	35
	TX	264436	1636	164	31
	TN	42092	278	28	30
	KS	82197	811	81	24
	OK	70003	667	67	24
	MS	47619	384	38	20
	IL	56299	643	64	18
	MO	69833	385	38	17
	IA	56258	550	55	16
	AL	51716	405	40	14
	KY	40320	208	21	13

Selected:6

19. Scroll to the left in the table until you see **STATE_NAME**. You see the names of six states highlighted in blue.
20. Take out your "Tornado Alley" worksheet.
21. Find Column 2: Highest number of strong or violent tornadoes.

(Q14) Write the names of these six states in Column 2 of your worksheet. (Some state names are filled in for you.)

M4 L1

22. Close the **Attribute Table**.

23. Click the **Clear All Selections** button.

Step 6: States with the highest concentration of tornadoes.

Looking at tornado density is a third way to decide which states are in Tornado Alley. This shows where the most tornadoes occured within the smallest area of land (closest together). When you map how close together or far apart tornadoes are, you call this a tornado "density" map.

1. Turn off **State_tornadoes**.
2. Turn on the **Tornado Density** layer.

The states cover up the Tornado Density layer, so you will change the symbol for the states.

3. Right-click the **States** layer name and choose **Properties**. This brings up the **Properties** window.

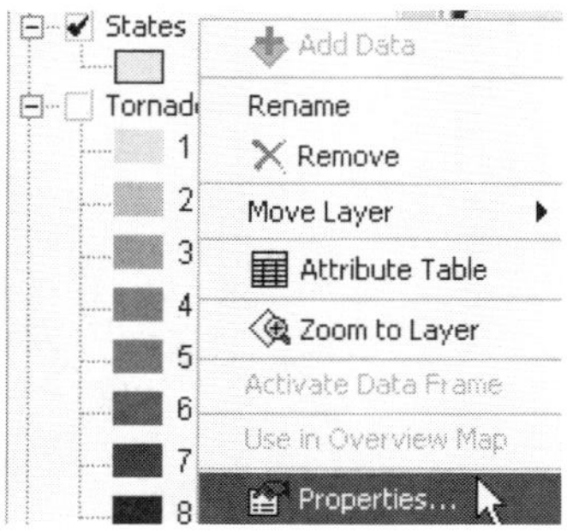

4. Change the **Style** to **Transparent fill**.

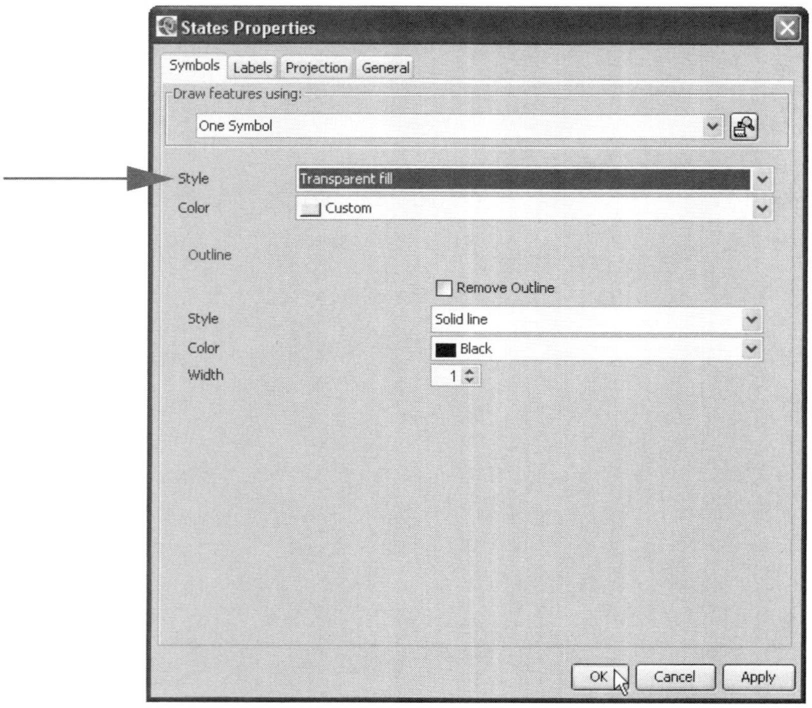

5. Click **OK**.

Now you can see the state outlines and labels on top of the Tornado Density layer.

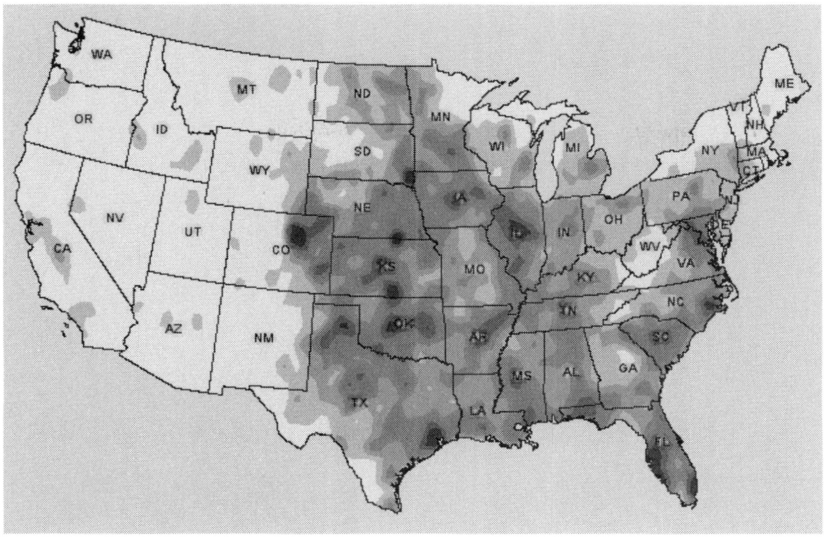

The dark red areas show where tornadoes are most dense (close together). The light red (pink) areas show where tornadoes are least dense (spaced far apart).

6. Click the **Tornado Density** layer name to highlight it. It becomes active.

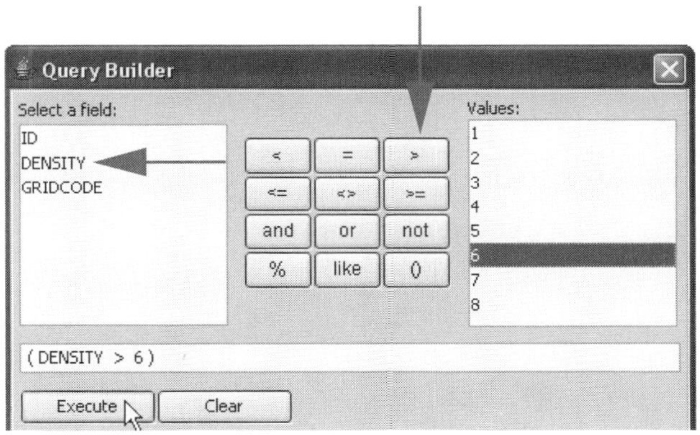

7. Click the **Query Builder** button.
8. Under **Select a field**, click **DENSITY**.
9. Click the > (greater than) button.
10. Under **Values**, click **6**.
11. Click **Execute**.
12. Close the **Query Builder** window.
13. Look at the map. The highlighted areas are where tornadoes are most dense (close together). You will use the GIS to draw a one-mile area (called a buffer) around each yellow area. Then you will use these buffers to select the states in the layer beneath them.

Buffer is a common GIS operation that draws an area of a certain distance around a feature. For instance, you might draw a buffer of a certain distance around a school to show a safety zone.

14. Click the **Buffer** button.
15. Click the box next to **Use buffer to select features from this layer**.

M4
L1

A list appears below the box.

16. Choose **States** from the list.

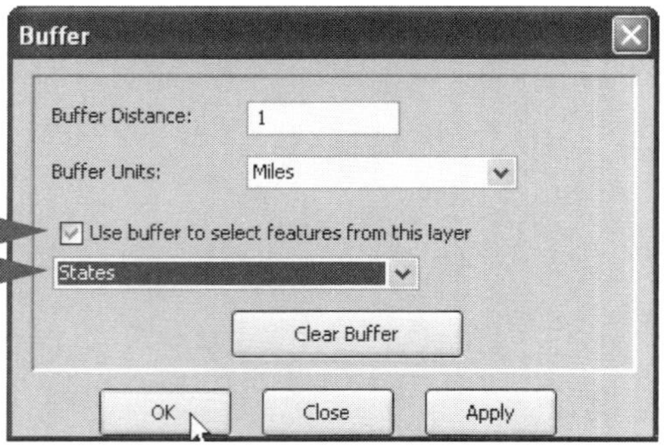

17. Click **OK**.

The states containing dense tornado areas are selected on the map.

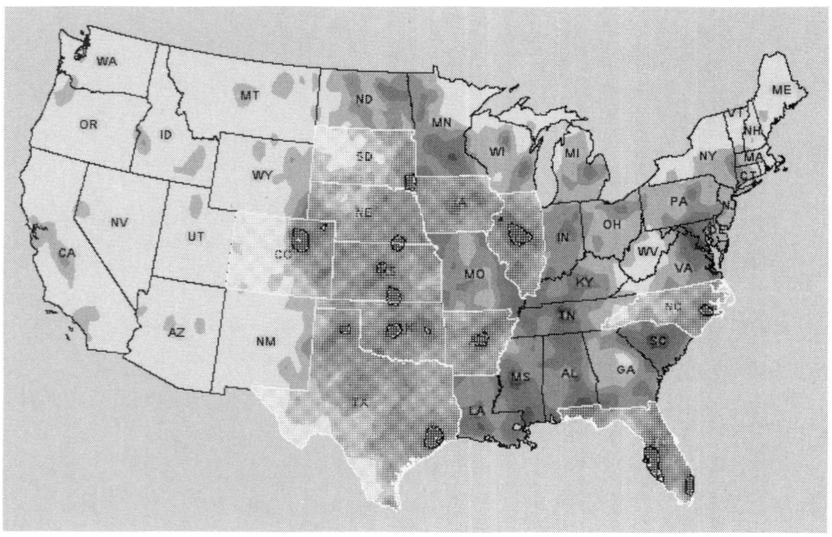

18. Click the **States** layer name to highlight it. It becomes active.

19. Click the **Attributes** button. The **Attributes of States** table opens.

20. Right-click **STATE_NAME**, and choose **Sort Selected Data to Top**.

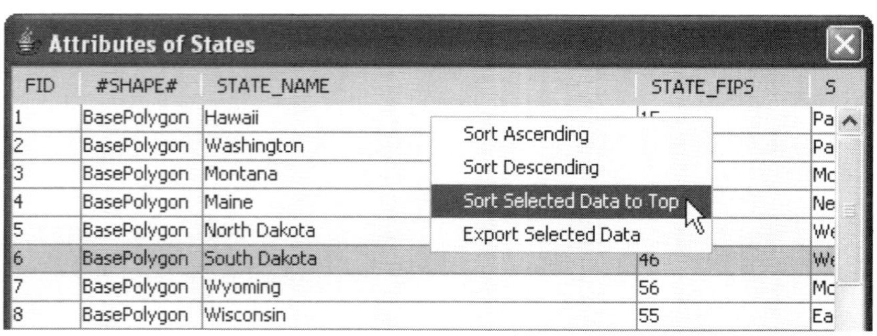

You see selected states highlighted in blue at the top of the table.

21. Look at the bottom of the table where it says **Selected**.

Q15 How many states contain dense tornado areas? ______________________________

22. Take out your "Tornado Alley" worksheet.
23. Find Column 3: Highest tornado density.

Q16 Write the names of the first six states in Column 3 of your "Tornado Alley" worksheet. (The rest of the state names are already filled in for you.)

24. Close the **Attribute Table**.

25. Click the **Clear All Selections** button.
26. Turn off the **Tornado Density** layer.

You have defined Tornado Alley in three different ways:
1. States where tornadoes occur most often
2. States where the strongest tornadoes occur
3. States where tornado density is highest

You listed these states on your "Tornado Alley" worksheet in Columns 1, 2, and 3.

27. Look at Columns 1, 2, and 3 now.

Q17 Which states appear in all three lists? Write the names of these states at the bottom of your "Tornado Alley" worksheet under "States in Tornado Alley." (Two state names have already been filled in for you.)

28. Take out your "Map of the 48 states" worksheet.
29. Color in these six states on the map.

Over 10 years, from 1995 to 2004, these states have been in Tornado Alley no matter what method you use to define this area.

Step 7: Where tornadoes occur during the seasons.

You have seen how tornadoes can occur close together over land. They can also occur close together over time. In this step you will look at where tornadoes occur during each season of the year and in each region of the United States.

1. Right-click the **States** layer name and choose **Properties**.
2. Under **Draw features using**, choose **Unique Symbols**.
3. For **Field for values**, scroll to the end of the list and choose **Region** (not SUB_REGION).
4. For **Color Scheme**, choose **Minerals**.
5. Click **OK**.

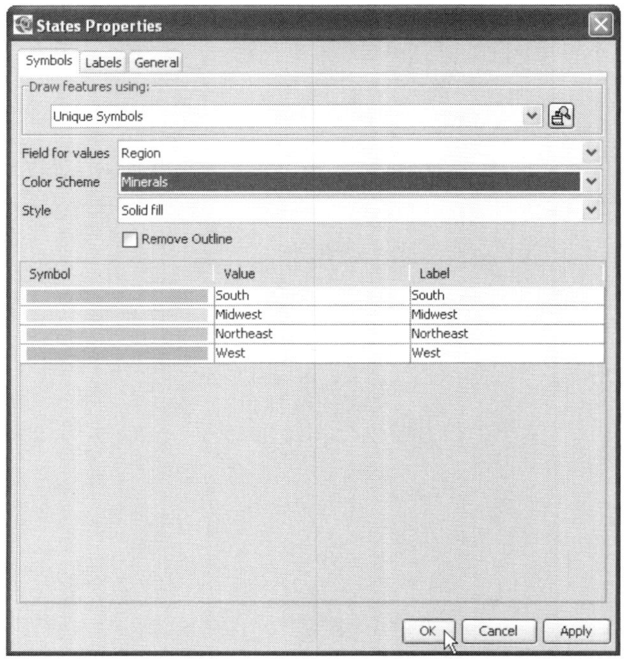

You see four United States regions, each with a different color, and you can see the state outlines.

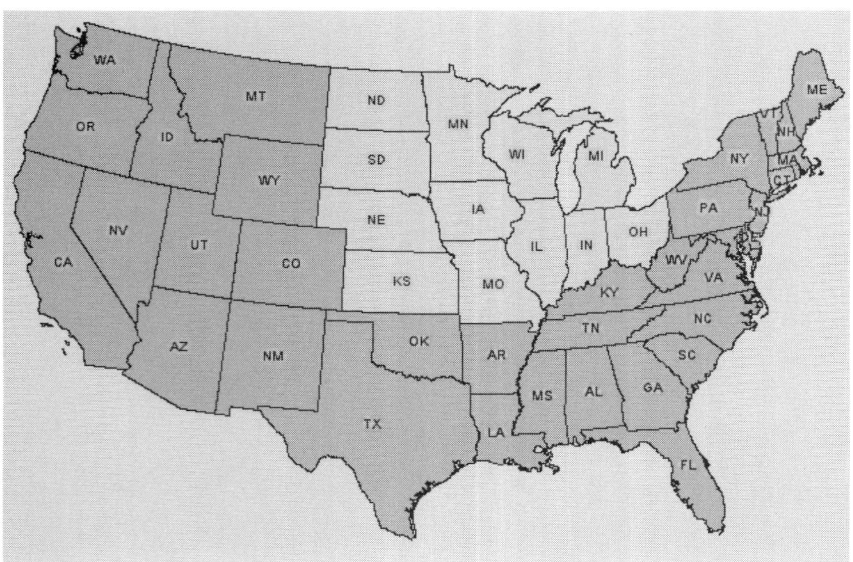

6. Turn on the **Winter_T** layer.

Winter tornadoes from 1995 to 2004 appear as green dots on the map.

Q18 **Look at the map. In what region of the country are most winter tornadoes? Hint: Look at the States legend to see the names and colors of the regions. (Circle the correct answer.)**

a. **Midwest**

b. **Northeast**

c. **South**

d. **West**

7. Click the **Winter_T** layer name to highlight it. It will become active.

8. Right-click the **Winter_T** layer name and choose **Attribute Table**.

The FID field contains a number for each winter tornado.

9. Scroll to the bottom of the table to see the highest FID number.

Q19 **How many tornadoes occurred in winter between 1995 and 2004?** ____________

10. Close the **Attribute Table**.
11. Turn off the **Winter_T** layer.
12. Turn on the **Spring_T** layer.

Spring tornadoes from 1995 to 2004 appear as red dots on the map.

Q20 **In what regions of the country are most spring tornadoes? (Circle the correct answer.)**

a. **Midwest**

b. **Northeast**

c. **South**

d. **West**

13. Click the **Spring_T** layer to make it active.

14. Right-click the **Spring_T** layer name and choose **Attribute Table**.

15. Scroll to the bottom of the table to see the highest FID number.

Q21 How many tornadoes occurred in spring between 1995 and 2004? ___________

16. Close the **Attribute Table**.

Q22 Look at the map. List two differences between winter tornadoes and spring tornadoes. (Hint: You can turn both layers on at the same time or turn them on and off to compare them.)

1. ____

2. ____

17. Turn off the **Spring_T** layer.
18. Turn on the **Summer_T** layer.

Summer tornadoes from 1995 to 2004 appear as yellow dots on the map.

Q23 In what region of the country are most summer tornadoes? (Circle the correct answer.)

a. Midwest

b. Northeast

c. South

d. West

Q24 Look at the map. List one difference between summer tornadoes and spring tornadoes. (Hint: You can turn both layers on at the same time or turn them on and off to compare them.)

____

19. Click the **Summer_T** layer to make it active.

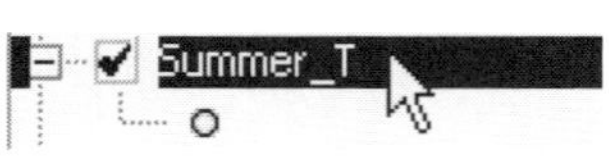

20. Right-click the **Summer_T** layer name and choose **Attribute Table.**
21. Scroll to the bottom of the table to see the highest FID number.

Q25 **How many tornadoes occurred in summer between 1995 and 2004?** ___________

22. Close the **Attribute Table**.
23. Turn off the **Summer_T** layer.
24. Turn on the **Fall_T** layer.

Fall tornadoes from 1995 to 2004 appear as blue dots on the map.

Q26 **In what region of the country are most fall tornadoes? (Circle the correct answer.)**

- **a. Midwest**
- **b. Northeast**
- **c. South**
- **d. West**

M4
L1

25. Click the **Fall_T** layer name to make it active.
26. Right-click the **Fall_T** layer name and choose **Attribute Table**.
27. Scroll to the bottom of the table to see the highest FID number.

Q27 **How many tornadoes occurred in fall between 1995 and 2004?** ______________

28. Close the **Attribute Table**.

Q28 **Which season had the most tornadoes?**

- **a. Winter**
- **b. Spring**
- **c. Summer**
- **d. Fall**

Q29 **What is the peak season for the southern states (the season when the South has the most tornadoes)?**

a. **Winter**

b. **Spring**

c. **Summer**

d. **Fall**

Q30 **What is the peak season for the northern states (the season when they have the most tornadoes)?**

a. **Winter**

b. **Spring**

c. **Summer**

d. **Fall**

Q31 **Discuss with your classmates some possible reasons that the locations of tornadoes change with the season. Write down two possible reasons.**

1. ____

2. ____

Step 8: Save your work and exit AEJEE.

1. Ask your teacher where and how to save your work.
2. Click the **File** menu and choose **Exit**.

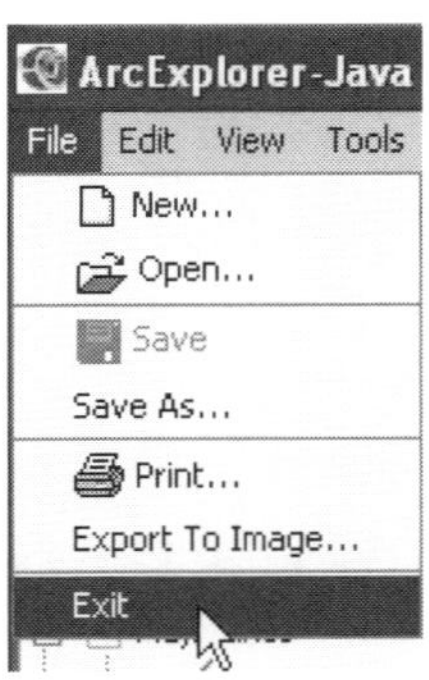

Conclusion

People who live in Tornado Alley are aware of the dangers of tornadoes and understand that they are part of life.

GIS can help researchers and scientists map patterns that show where tornadoes touch down and study how tornadoes behave on the ground, including how far and fast they move and how wide an area they cover. Their research leads to better warning systems and better methods for predicting when a tornado might occur. These improvements make it safer for people who live in Tornado Alley.

M4
L1

Name: ______________________

Date: ______________________

Module 4, Lesson 1

Finding Tornado Alley

M4 L1

Worksheet 1: Tornado Alley

Column 1	Column 2	Column 3
Highest average number of tornadoes	Highest number of strong or violent tornadoes	Highest tornado density
1.	1.	1.
2.	2.	2.
3.	3.	3.
4.	4.	4.
5.	5.	5.
6.	6.	6.
7. Iowa	7. Illinois	7. Oklahoma
8. Colorado	8. Missouri	8. North Carolina
9. Minnesota	9. Iowa	9. Texas
10. Arkansas	10. Alabama	10. Arkansas
11. Alabama	11. Kentucky	11. Florida

States in Tornado Alley

1. Texas
2. ______________________
3. ______________________
4. ______________________
5. ______________________
6. Arkansas

Worksheet 2: Map of the 48 states

Name: ____________________ Date: ____________

Name: ______________________________

Date: ______________________________

Module 4, Lesson 2

Analyzing historical tornadoes

Meteorologists study storms that produce tornadoes. They use tools such as forecasting (predicting the weather), research (observing and trying to understand the weather), and storm chasing (following and watching storms close-up). Some tornadoes and tornado outbreaks may stick even in your memory. Some students grow up to study the science of tornadoes because of a personal experience or out of curiosity about a particular outbreak. Perhaps you will decide to be a meteorologist yourself one day!

In this GIS activity, you will see how tornadoes are classified into weak, strong, and violent categories. You will also take a close look at some of the memorable tornado outbreaks in history.

Step 1: Start AEJEE.

1. Ask your teacher how to start the AEJEE software.
2. Click the **Maximize** button at the top of the AEJEE window. AEJEE will now take up all the space on your screen.

Step 2: Open the project.

1. Click the **Open** button.
2. Navigate to your **OurWorld1\Module4** folder.
3. Choose **Historical.axl**.
4. Click **Open**.

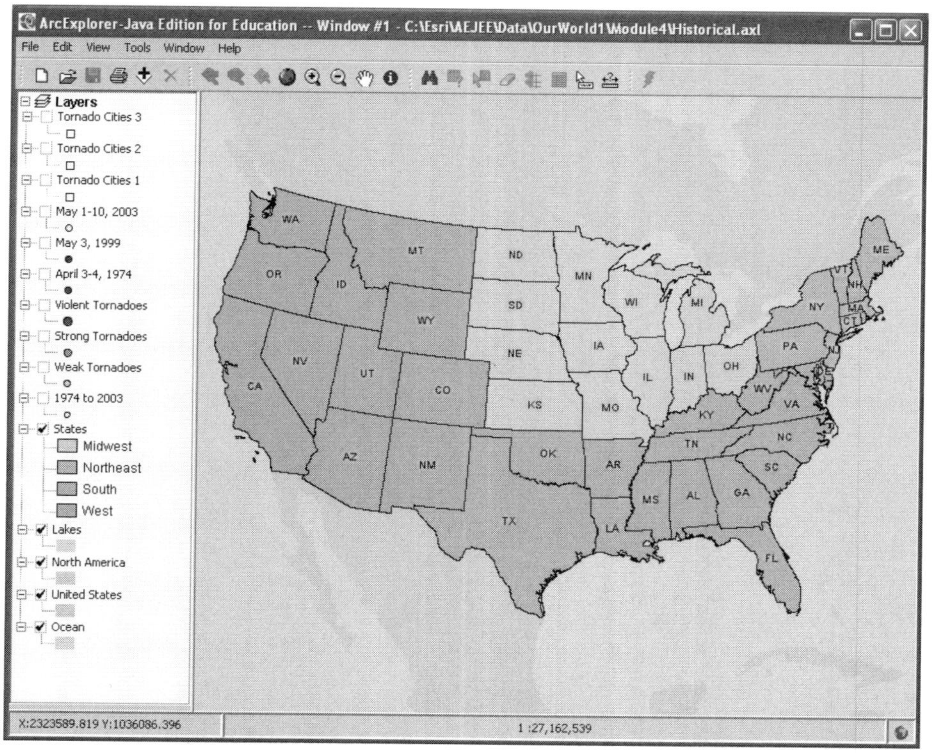

You see states in four regions of the United States. The States legend shows you the name of each region.

Step 3: Thirty years of tornadoes.

From 1974 to 2003, the United States had three memorable tornado outbreaks. A tornado outbreak occurs when many tornadoes form from a single storm. Before looking at these particular outbreaks, you will look at the tornadoes in this 30-year time period.

1. Turn on the **1974 to 2003** layer by clicking the box next to the layer name.

Q1 **Which region had the fewest tornadoes between 1974 and 2003?**

- **a Midwest**
- **b. Northeast**
- **c. South**
- **d. West**

2. Click the **1974 to 2003** layer name to highlight it. It becomes active.

3. Click the **Zoom to Active Layer** button.

Now you see all the states, including Alaska and Hawaii.

Q2 **Which state had no tornadoes between 1974 and 2003?** ____________________

4. Click the **Previous Extent** button to zoom back to the 48 states.
5. Right-click the **1974 to 2003** layer name and choose **Attribute Table**. (Be patient while the table opens.)

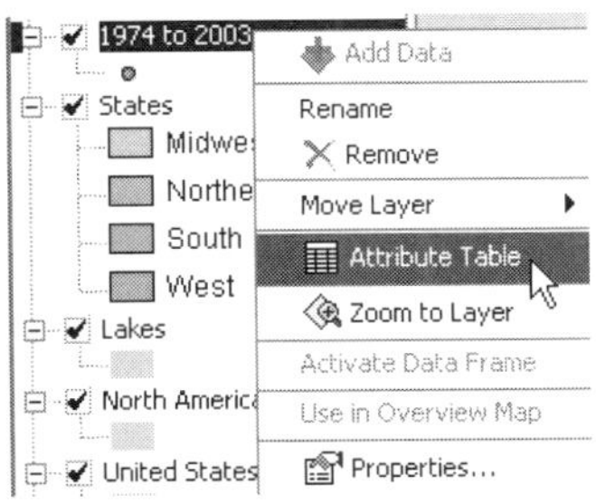

The first field in the table, called FID (which stands for feature identification number), contains a unique number for every tornado.

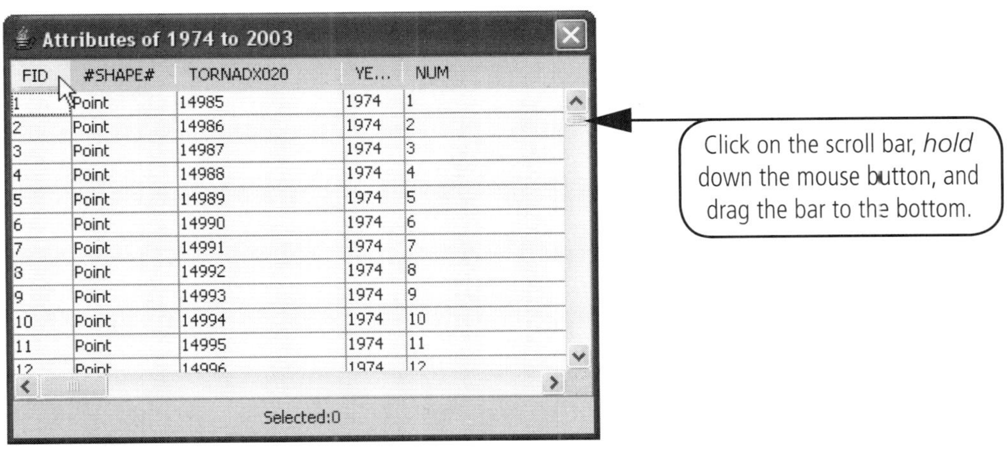

FID	#SHAPE#	TORNADX020	YE...	NUM
1	Point	14985	1974	1
2	Point	14986	1974	2
3	Point	14987	1974	3
4	Point	14988	1974	4
5	Point	14989	1974	5
6	Point	14990	1974	6
7	Point	14991	1974	7
8	Point	14992	1974	8
9	Point	14993	1974	9
10	Point	14994	1974	10
11	Point	14995	1974	11
12	Point	14996	1974	12

6. Scroll to the bottom of the table.
7. Look at the number in the FID field.

Q3 **How many tornadoes occurred between 1974 and 2003? (Write the number here. You will use this number in the next step.)** ____________________

8. Close the **Attribute Table**.

Step 4: Weak tornadoes.

A tornado is like a spinning wind tunnel. It extends from the thundercloud in the sky to the ground below. (To be a tornado, it must touch the ground.) The strength of a tornado is measured by the speed of the spinning wind and the amount of damage to things on the ground—houses, cars, trains, trees, and roads. Tornado strength is measured on a scale called the Fujita Tornado Damage Scale, or F-Scale for short.

1. Take out the "Tornado F-scale" handout that your teacher gave you. It shows six different strengths, from the weakest (F0) to the most violent (F5). For each size, you see the wind speed and a description of the damage.

   Tornadoes may also be described as weak, strong, or violent:
   - F0 and F1 tornadoes are considered weak.
   - F2 and F3 tornadoes are considered strong.
   - F4 and F5 tornadoes are considered violent.

2. Return to the map on your computer screen.
3. Turn off the **1974 to 2003** layer by clicking the check mark next to the layer name.

4. Turn on **Weak Tornadoes** by clicking the box next to the layer name.

You see all the weak (F0 and F1) tornadoes that occurred between 1974 and 2003. Weak tornadoes cause very few deaths. Weak tornadoes can last from about 1 minute to 10 minutes. Their winds are from 40 to 112 miles per hour.

(Q4) Which region had the fewest weak tornadoes between 1974 and 2003?

a. Midwest

b. Northeast

c. South

d. West

(Q5) Look at your "Tornado F-scale" handout. Write down two examples of damage that occurs with F0 tornadoes.

1. ______________________________

2. ______________________________

(Q6) **Look at your "Tornado F-scale" handout again. Write down two examples of damage that occurs with F1 tornadoes.**

1. ______________________________

2. ______________________________

5. Right-click the **Weak Tornadoes** layer name and choose **Attribute Table**.

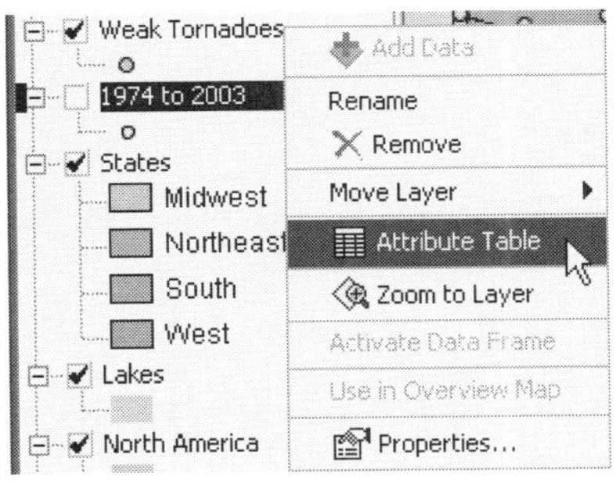

The first field in the table is the FID (feature identification) field. It contains a unique number for each tornado. The highest FID number tells you the total number of tornadoes in this layer.

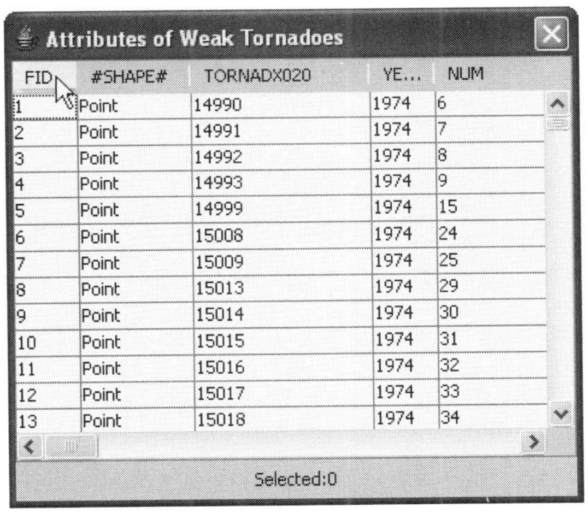

6. Scroll to the bottom of the table.

(Q7) **How many weak tornadoes occurred in the United States between 1974 and 2003?** ______________________________

Q8 **What percentage of all tornadoes were weak? To answer this question, you will divide the number of weak tornadoes by the total number of tornadoes (you wrote this number down in Q3). Then you will multiply the answer by 100. You may need to use a calculator.**

Weak tornadoes ÷ total tornadoes = 0.__ __ X 100 = __ __ percent (without the decimal)

7. Close the **Attribute Table**.

Step 5: Strong tornadoes.

Strong tornadoes are responsible for nearly one-third of all the deaths caused by tornadoes. Strong tornadoes can last 20 minutes or longer. Their winds are from 113 to 206 miles per hour.

How do the wind speeds of strong tornadoes compare to other speeds?

- A cheetah, the fastest land animal, can run as fast as 70 miles per hour.
- The fastest roller coaster in the world can go 120 miles per hour.
- The fastest road car can reach speeds of 240 miles per hour.
- The fastest NASA jet plane can travel up to 2,274 miles per hour.

1. Turn off the **Weak Tornadoes** layer.

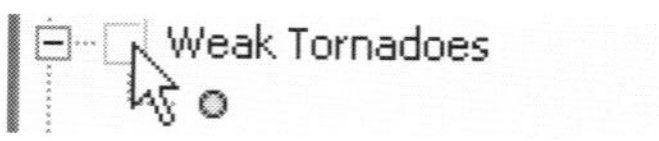

2. Turn on the **Strong Tornadoes** layer.

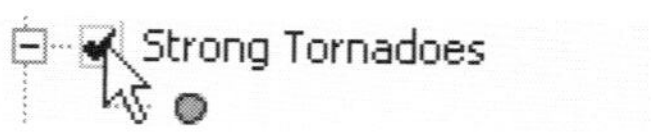

You see all the strong (F2 and F3) tornadoes that occurred from 1974 to 2003.

Q9 **Which region had the fewest strong tornadoes between 1974 and 2003?**

a Midwest

b. Northeast

c. South

d. West

Q10 **Look at the map. Does it look like there were more strong tornadoes or weak tornadoes from 1974 to 2003? (Hint: Turn the layers on and off to compare them.)**

a. **More strong tornadoes**

b. **More weak tornadoes**

Q11 **Look at your "Tornado F-scale" handout. Write two examples of damage that occurs with F2 tornadoes.**

1. ______________________________

2. ______________________________

Q12 **Look at your "Tornado F-scale" handout. Write two examples of damage that occurs with F3 tornadoes.**

1. ______________________________

2. ______________________________

M4 L2

3. Right-click the **Strong Tornadoes** layer name and choose **Attribute Table**.

The FID (feature identification) field contains a unique number for each tornado. The highest number tells you the total number of tornadoes in this layer.

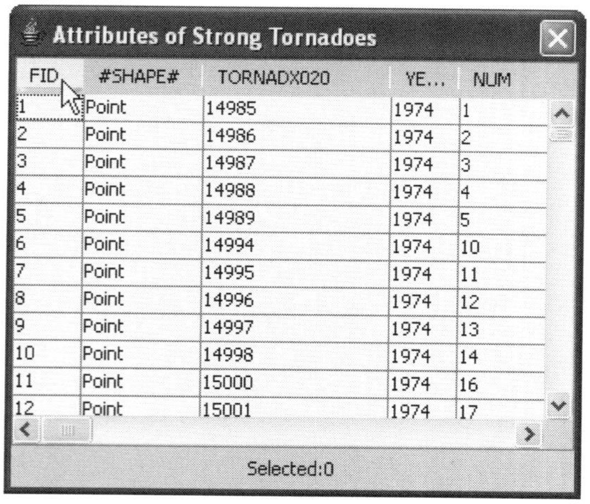
Attributes of Strong Tornadoes

FID	#SHAPE#	TORNADX020	YE...	NUM
1	Point	14985	1974	1
2	Point	14986	1974	2
3	Point	14987	1974	3
4	Point	14988	1974	4
5	Point	14989	1974	5
6	Point	14994	1974	10
7	Point	14995	1974	11
8	Point	14996	1974	12
9	Point	14997	1974	13
10	Point	14998	1974	14
11	Point	15000	1974	16
12	Point	15001	1974	17

Selected:0

4. Scroll to the bottom of the table.

Q13 **How many strong tornadoes occurred between 1974 and 2003? ______________**

Q14 **What percentage of all tornadoes were strong? To answer this question, you will divide the number of strong tornadoes by the total number of tornadoes (you wrote this number down in Q3). Then, you will multiply the answer by 100. You may need to use a calculator.**

Strong tornadoes ÷ total tornadoes = 0.__ __ X 100 = __ __ percent (without the decimal)

5. Close the **Attribute Table**.

Step 6: Violent tornadoes.

Violent tornadoes are responsible for more than two-thirds of all the deaths caused by tornadoes. A violent tornado can last more than one hour. The winds from a violent tornado range from 207 to 318 miles per hour. The most violent tornadoes create the strongest known winds on earth!

1. Turn off **Strong Tornadoes**.

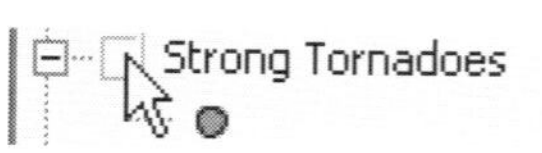

2. Turn on **Violent Tornadoes**.

You see all the violent (F4 and F5) tornadoes that occurred between 1974 and 2003.

Q15 What region had the fewest violent tornadoes between 1974 and 2003?

a. Midwest

b. Northeast

c. South

d. West

Q16 Which regions had the most violent tornadoes between 1974 and 2003? (Circle all the correct answers.)

a. Midwest

b. Northeast

c. South

d West

Q17 Were there more violent tornadoes or strong tornadoes from 1974 to 2003? (Hint: Turn the layers on and off to compare them.)

a. More violent tornadoes

b. More strong tornadoes

Q18 Look at your "Tornado F-scale" handout. Write two examples of damage that occurs with F4 tornadoes.

1. ______________________________

2. ______________________________

Q19 Look at your "Tornado F-scale" handout. Write two examples of damage that occurs with F5 tornadoes.

1. ______________________________

2. ______________________________

3. Right-click the **Violent Tornadoes** layer name and choose **Attribute Table**.

   The FID field contains a unique number for each tornado. The highest number tells you the total number of tornadoes in this layer.

4. Scroll to the bottom of the table.

(Q20) How many violent tornadoes occurred between 1974 and 2003? ______________

(Q21) What percentage of all tornadoes were violent? You already calculated the percentage of weak and strong tornadoes. Write those down here.

Percentage of weak tornadoes: __ __percent (from Q8)

Percentage of strong tornadoes: __ __percent (from Q14)

Add these two numbers together: __ __percent
Subtract this number from 100%.

What is the percentage of violent tornadoes?

a. 1 percent

b. 2 percent

c. 3 percent

d. 4 percent

5. Close the **Attribute Table**.

Step 7: The most violent tornado outbreak that we know about in history happened on April 3–4, 1974.

1. Turn off the **Violent Tornadoes** layer.
2. Turn on the **April 3–4, 1974** layer.

Your map shows all the tornadoes that took place in these two days.

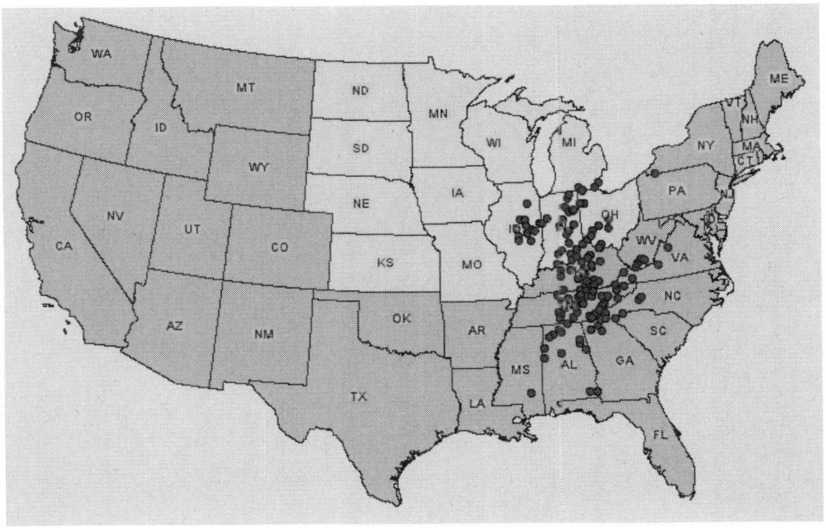

3. Turn on the **Tornado Cities 1** layer. You see the city of Xenia, Ohio, where the worst tornado took place.
4. Right-click the **Tornado Cities 1** layer name and choose **Attribute Table**. You see one record (row) in the table.

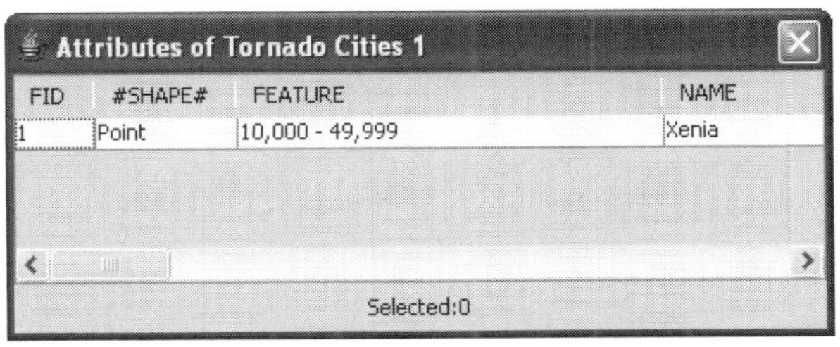
Attributes of Tornado Cities 1

FID	#SHAPE#	FEATURE	NAME
1	Point	10,000 - 49,999	Xenia

Selected:0

5. Scroll to the right in the table until you see the **POP_2000** field. This field contains the population for the year 2000.

Q22 **How many people lived in Xenia in 2000?** ______________________

6. Close the **Attribute Table**.

The tornado that hit Xenia was an F5. It destroyed half the town and caused 100 million dollars in property damage.

7. Take out the "Historical tornadoes" worksheet that your teacher gave you.
8. Look at the column called **EVENT 1**. There is a lot of information about the April 3–4, 1974, tornado outbreak. Use this information to answer the questions below.

Q23 **How many tornadoes occurred during this 24-hour period?** ______________

Q24 **How many states were affected?** ______________________________

Q25 **How many F5 tornadoes occurred on April 3 and 4?** ______________________________

The total number of injuries and deaths is not filled in. You will use the GIS to get this information.

9. Turn off the **Tornado Cities 1** layer.
10. Click the **April 3-4, 1974** layer name to make it active. It becomes highlighted.

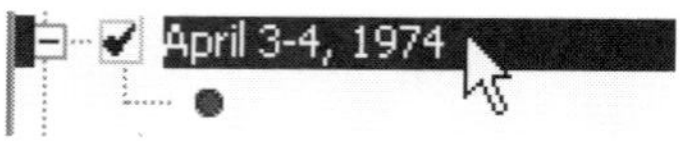

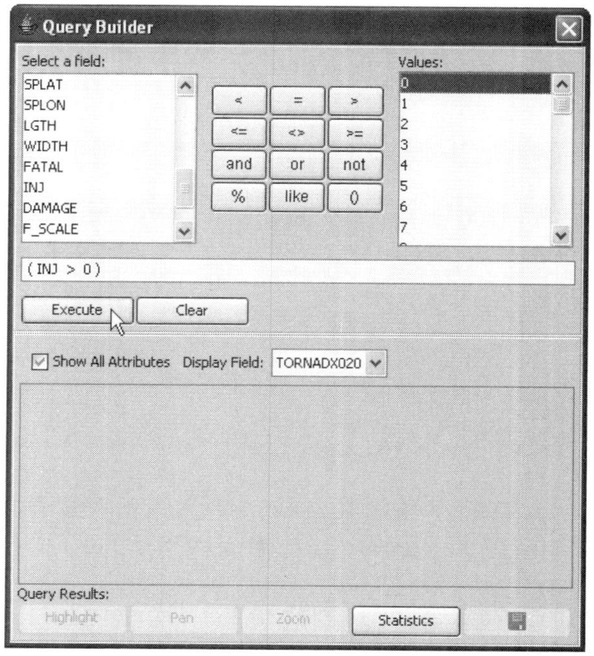

11. Click the **Query Builder** button.
12. Under **Select a field**, scroll down and click **INJ**.
13. Click the > (greater than) button.
14. Under **Values**, click **0** (the first value in the list).
15. Click **Execute**.

    You selected all the tornadoes that had more than 0 (zero) injuries.

Q26 **How many tornadoes caused injuries? (Hint: Look at the Query Results at the bottom of the window.)** ______________________________

16. Click the **Statistics** button.
17. Under **Select a field**, scroll down and click **INJ**.
18. Click the box next to **Use Query Results?**

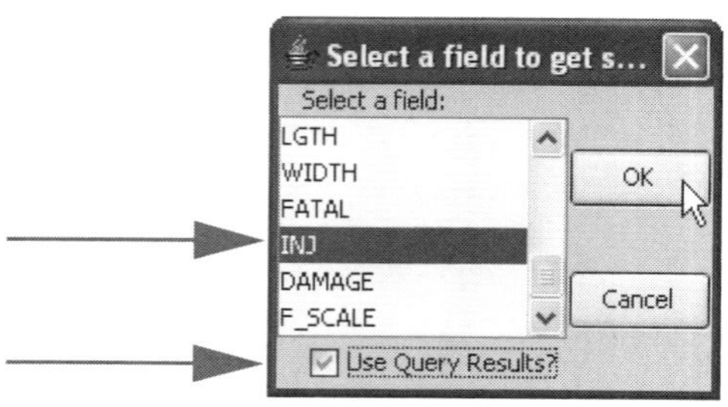

19. Click **OK**. A window called "Statistics Results for field:INJ" opens.
20. Look at the **Statistics Results** window.
21. Look at the bottom row that says **Total**.

Q27 **What is the total number of injuries that occurred? Write this number on your "Historical tornadoes" worksheet.**

22. Close the **Statistics Results** window.
23. In the **Query Builder** window, click **Clear**.

    Now you will find out the total number of deaths that occurred.

24. Under **Select a field**, click **FATAL**.
25. Click the > (greater than) button.
26. Under **Values**, click **0** (the first value in the list).
27. Click **Execute**.

    You selected all the tornadoes that had more than 0 (zero) deaths.

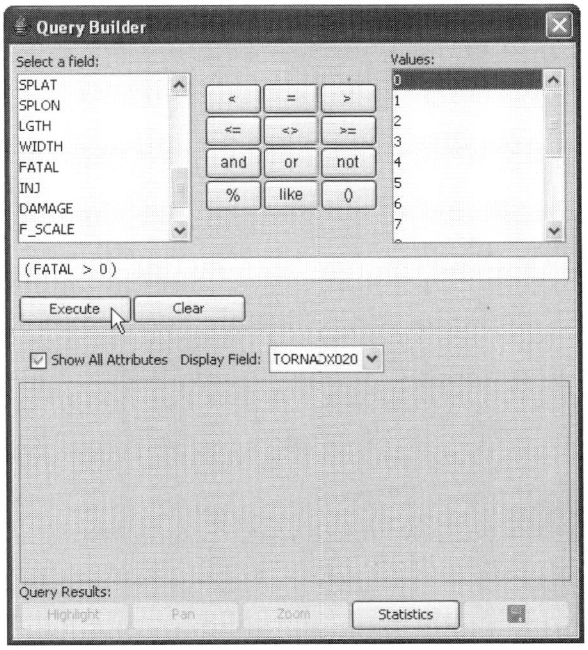

M4
L2

Q28 **How many tornadoes caused deaths? (Hint: Look at the Query Results at the bottom of the window.)** ______________________

28. Click the **Statistics** button.
29. Under **Select a field**, click **FATAL**.
30. Click the box next to **Use Query Results?**

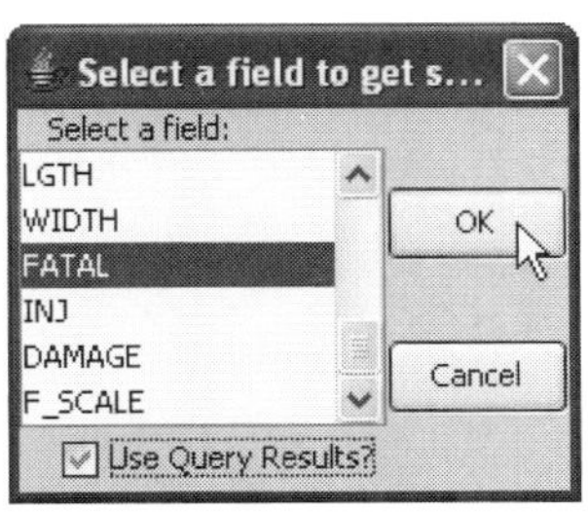

31. Click **OK**. A window called "Statistics Results for field: FATAL" opens.

32. Look at the **Statistics Results** window.
33. Look at the bottom row that says **Total**.

Q29 **What is the total number of deaths that occurred? Write this number on your "Historical tornadoes" worksheet.**

34. Close the **Statistics** window.
35. Close the **Query Builder**.
36. Click the **Clear All Selections** button.

Now you have filled in the missing information for Event 1.

Q30 **Look at your map, and look at your "Historical tornadoes" worksheet. Why do you think the tornado outbreak of April 3–4, 1974, is considered the most violent tornado outbreak in history? List two reasons.**

1. ____

2. ____

Notice that the total damage for this tornado outbreak was more than 600 million dollars. This is the amount it costs to launch the space shuttle!

Step 8: The most costly tornado of all time occurred on May 3, 1999.

1. Turn off the **April 3–4, 1974** layer.
2. Turn on the **May 3, 1999** layer.

Your map shows 71 tornadoes that took place on that day.

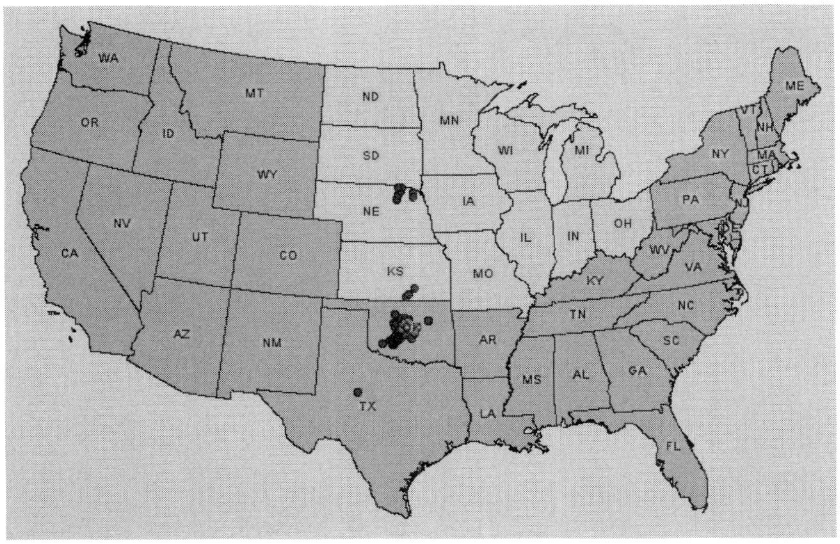

3. Turn on the **Tornado Cities 2** layer. You see the city of Moore, a suburb of Oklahoma City, Oklahoma. This is where the worst tornado took place.
4. Right-click the **Tornado Cities 2** layer name and choose **Attribute Table**. There is one record (row) in the table.

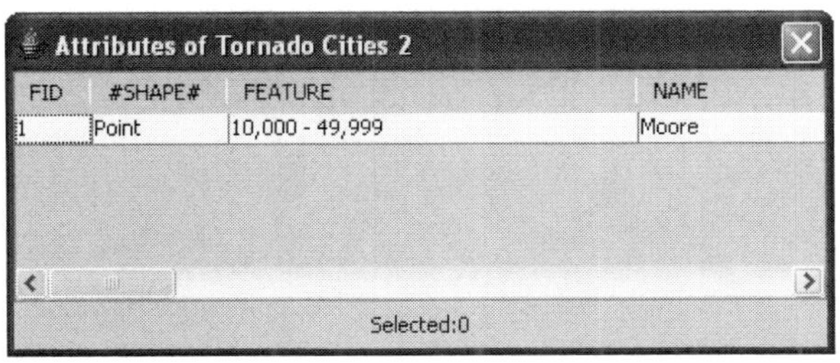
Attributes of Tornado Cities 2

FID	#SHAPE#	FEATURE	NAME
1	Point	10,000 - 49,999	Moore

Selected:0

5. Scroll all the way to the right until you see the **HSE_UNITS** field. This field contains the number of housing units (houses).

Q31 **How many housing units (houses) are in Moore?** ______________________________

6. Close the **Attribute Table**.

The tornado that hit Moore destroyed more than 3,000 of these houses and caused one billion dollars in damage.

7. Right-click the **May 3, 1999** layer name and choose **Attribute Table**.
8. Scroll to the **F_SCALE** field.
9. Right-click **F_SCALE** and choose **Sort Descending**. This reorders the tornadoes from the highest F-scale to the lowest F-scale.

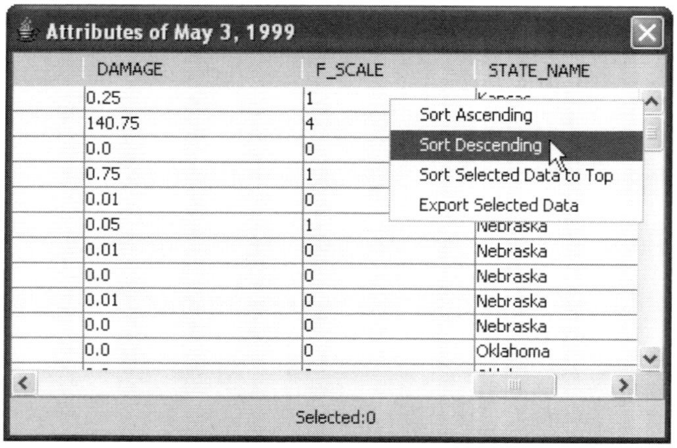

(Q32) **What is the F-scale of the strongest tornado on this day?** ____________________

10. Click the first record to highlight it. It is also highlighted on the map.

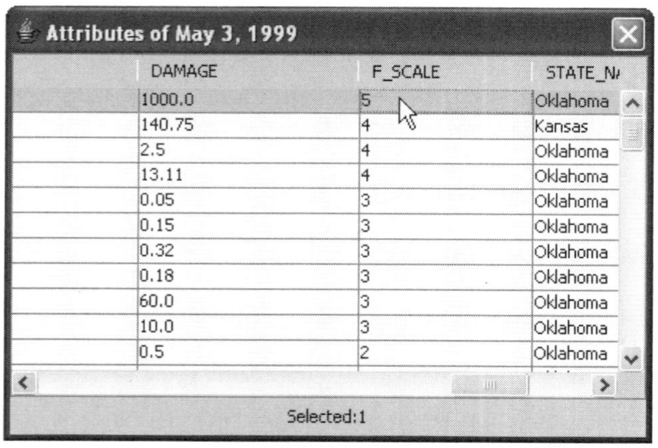

11. Look at the map on your screen.

> **Note:** If the attribute table is covering your map, place your mouse on the words "Attributes of May 3, 1999" at the top of the window. Hold down your mouse button and drag the window out of the way.

You see a yellow dot representing the F5 tornado very close to a white square representing the suburb of Moore. This F5 tornado hit the suburb of Moore on May 3, 1999.

While moving through the area, this F5 tornado created a visible path called a track.

- You can estimate how far this tornado traveled by measuring the length of its track.
- You can estimate the width of this tornado by measuring the width of its track.

12. In the Attribute Table, scroll back until you see the **LGTH** field.

    The numbers in this field represent tenths of a mile. That means that you need to divide the number in the table by 10 to get the actual length (Example: 146 ÷ 10 = 14.6 miles).

Q33 **How far did this F5 tornado travel? Divide the answer by 10.**
Length value in table = __ __ __ ÷ 10 = __ __.__ miles

13. In the Attribute Table, right-click **LGTH** and choose **Sort Descending**. This reorders the tornadoes from the longest track to the shortest.

Q34 **Did this F5 (highlighted) tornado have the longest track of any tornado that occurred that day (May 3, 1999)?**

a. Yes

b. No

M4
L2

14. In the Attribute Table, find the **WIDTH** field (it is right next to the **LGTH** field).

    The numbers in this field represent tens of feet. That means that you need to multiply the number in the table by 10 to get the actual length (Example: 528 x 10 = 5,280 feet or 1 mile). Tornadoes can be more than a mile wide.

Q35 **What was the width of the F5 tornado? Multiply the answer by 10. (You can ignore the number after the decimal.)**
Width value in table = __ __ __ X 10 = __ __ __ __ feet

15. In the Attribute Table, right-click **WIDTH** and choose **Sort Descending**. This reorders the tornadoes from the widest to the narrowest.

Q36 **Was this F5 (highlighted) tornado the widest tornado that occurred that day?**

a. Yes

b. No

Q37 **How many tornadoes were wider?** ________________________

16. Close the **Attribute Table**.
17. Click the **May 3, 1999** layer name to highlight it. It becomes active.

18. Click the **Clear All Selections** button.
19. Look at your "Historical tornadoes" worksheet.
20. Look under the column called **EVENT2**.

You see many details of the May 3, 1999, tornado outbreak.

Notice that the number of states, the names of states, and the regions affected are not filled in. You will use the GIS to get this information.

21. Turn off the **Tornado Cities 2** layer.
22. Click the **Select Features** tool and choose **Rectangle**.

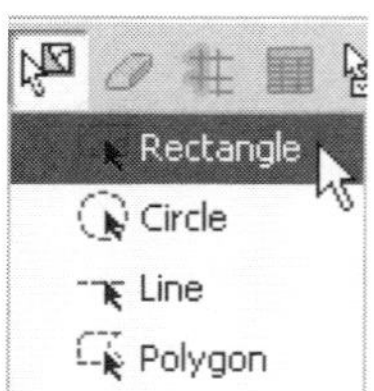

You will use this tool to select all the tornadoes.

23. Click anywhere in the state of Montana (MT) and *hold* down the mouse button.
24. Drag the mouse down and to the right until you reach the state of Louisiana (LA).

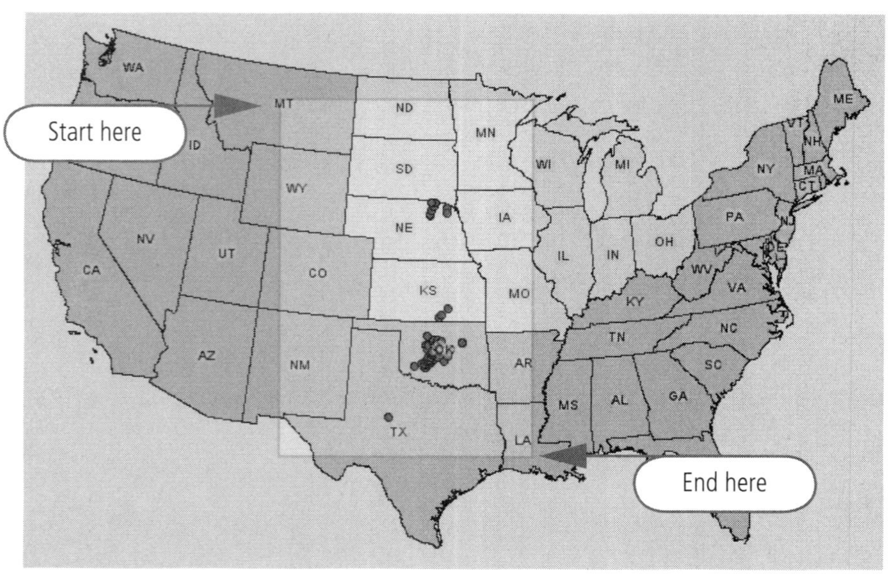

25. Let go of the mouse button.

All the tornadoes in the May 3, 1999 layer should be highlighted yellow.

> **Note:** If you make a mistake, click the **Clear All Selections** button and try again.

Now you will select all the states that these tornadoes are in.

26. Click the **Buffer** tool. The Buffer tool creates a one-mile shape around each selected tornado. You will use these buffers to select the states in the layer beneath the tornadoes.
27. Click the box next to **Use buffer to select features from this layer**. A list appears below the box.
28. Select **States** from the list.

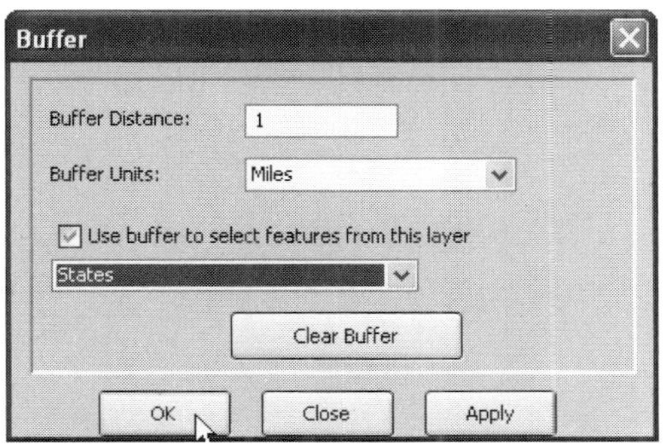

29. Click **OK**.

The GIS uses the buffer (shape) around each tornado to select the state that it is in. The selected states are highlighted in yellow on the map.

30. Click the **States** layer name to highlight it (make it active).

31. Click the **Attributes** button. The Attribute Table for States opens.
32. Look at the bottom of the table next to **Selected**.

Q38 **How many states were hit by tornadoes on May 3, 1999? Write your answer on your "Historical tornadoes" worksheet.**

33. In the table, right-click the **STATE_NAME** field and choose **Sort Selected Data to Top**.

M4
L2

(Q39) Which states were hit by tornadoes on May 3, 1999? Write the names of the states on your "Historical tornadoes" worksheet.

34. Close the **Attribute Table**.

(Q40) Look at the legend for the States layer. The highlighted states are located in which regions? Write your answer on your "Historical tornadoes" worksheet.

35. Click the **Clear All Selections** button.

Now you have filled in the missing information for Event 2.

You can see on your worksheet that the total damage for the May 3, 1999, tornado outbreak was 1.6 billion dollars. The single F5 tornado that hit the Oklahoma City suburb of Moore caused one billion dollars in damage all by itself—it is the most expensive tornado of all time.

Step 9: More tornadoes occurred between May 1 and May 10, 2003, than any other ten-day period since record keeping began.

1. Turn off the **May 3, 1999** layer.
2. Turn on the **May 1-10, 2003** layer.

Your map shows nearly 400 tornadoes that hit the central and southern United States from May 1-10, 2003.

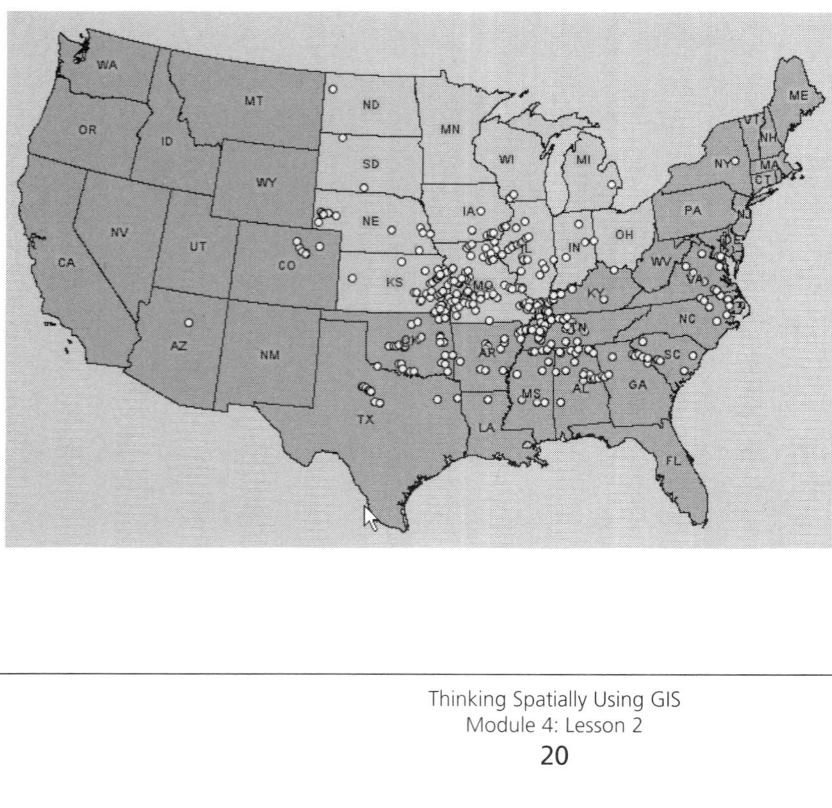

Q41 **Look at your map. Describe in your own words how this tornado outbreak looks different from the other two outbreaks you have already looked at. (Hint: Turn the layers on and off to compare them. How far do the tornadoes extend in all directions? Which regions and states were hit?)**

____

____

____

3. Take out your "Historical tornadoes" worksheet.
4. Look at the column called **EVENT 3**.

You see information about the tornado outbreak of May 1 to May 10, 2003.

On your worksheet, the F-scale and total number of strong and violent tornadoes information is missing. You will use the GIS to get this information.

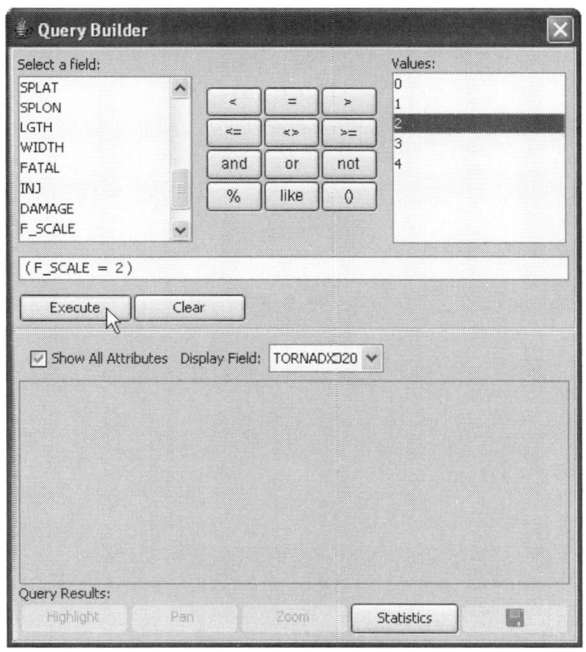

5. Click the **May 1-10, 2003** layer name to highlight it. It becomes active.

6. Click the **Query Builder** button.
7. Under **Select a field**, choose **F_SCALE**.
8. Click the = (equals) button.
9. Under **Values**, click **2**.
10. Click **Execute**.
11. Look at the **Query Results** at the bottom of the window.

Q42 **How many F2 tornadoes occurred between May 1 and May 10, 2003? Write the number on your "Historical tornadoes" worksheet.**

Now you will follow the same steps to find out the number of F3 tornadoes.

12. In the equation box, delete the number **2** and type **3**.

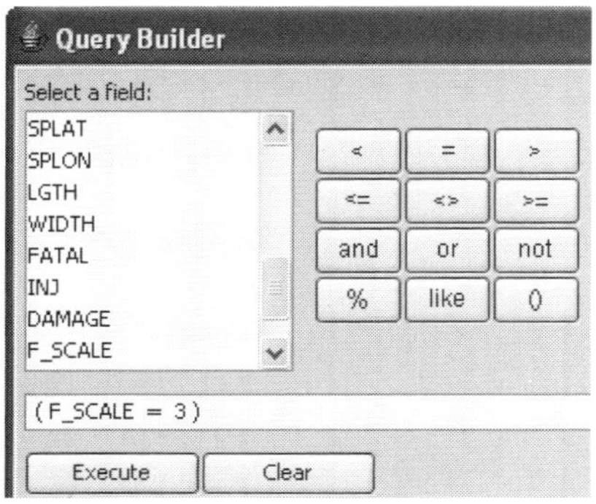

13. Click **Execute**.
14. Look at the **Query Results** at the bottom of the window.

Q43 How many F3 tornadoes occurred between May 1 and May 10, 2003? Write the number on your "Historical tornadoes" worksheet.

You will follow the same steps to find out the number of F4 tornadoes.

15. In the equation box, delete the number **3** and type **4**.
16. Click **Execute**.

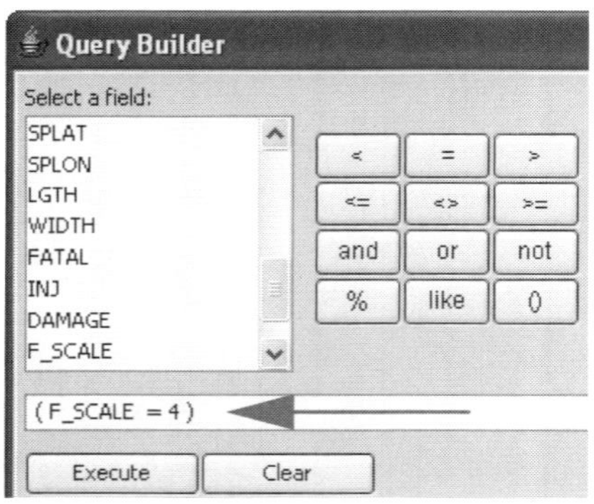

17. Look at the **Query Results** at the bottom of the window.

Q44 How many F4 tornadoes occurred from May 1-10, 2003? Write the number on your "Historical tornadoes" worksheet.

18. There were no F5 tornadoes that took place, so write "**0**" on your handout.

Now that you know the number of F2, F3, F4, and F5 tornadoes, you can use simple addition to find out the number of strong and violent tornadoes. You learned that strong tornadoes have an F-scale of 2 or 3, and violent tornadoes have an F-scale of 4 or 5.

Q45 **Add the number of F2 and F3 tornadoes together.**
__ __ (number of F2s) + __ __ (number of F3s) = __ __ Strong tornadoes. Write this number on your "Historical tornadoes" worksheet.

Q46 **Add the number of F4 and F5 tornadoes together.**
__ __ (number of F4s) + __ __ (number of F5s) = __ __ Violent tornadoes. Write this number on your "Historical tornadoes" worksheet.

19. In the **Query Builder** window, click **Clear**.
20. Close the **Query Builder**.

You have filled in the missing information for Event 3.

Your worksheet shows that the total damage for this tornado outbreak was 3.4 billion dollars, more than the previous two events combined.

Q47 **Look at your "Historical tornadoes" worksheet. Why do you think the 2003 tornado outbreak caused more damage than the other two outbreaks? Discuss this with your classmates, and write your answer here.**

____

____

____

Step 10: Save your work and exit AEJEE.

1. Ask your teacher where and how to save your work.
2. Click the **File menu** and choose **Exit**.

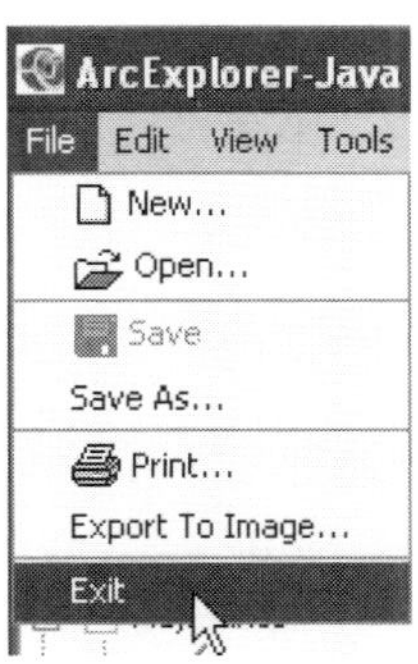

Conclusion

Tornadoes and other natural disasters are events that affect many people in the United States. People are working together to create more effective warning and tracking systems for these disasters in hopes of saving lives.

GIS is one of many tools that help researcher and scientists make educated and effective decisions on predicting when and where these weather patterns will form as well as assist in the recovery process after a disaster has struck.

Name: ______________________________

Date: ______________________________

Module 4, Lesson 2

Analyzing historical tornadoes

Worksheet: Historical tornadoes

	Event 1	Event 2	Event 3
Date	April 3–4, 1974	May 3, 1999	May 1–10, 2003
Why is it remembered?	Largest outbreak	Most expensive single tornado	More tornadoes than any other 10-day period
Total number of tornadoes	148 tornadoes	71 tornadoes	401–412 tornadoes
Cities most affected	Xenia, Ohio	Moore, Oklahoma, (suburb of Oklahoma City)	Kansas City, Kansas; Oklahoma City, Oklahoma; Pierce City, Missouri; Jackson, Tennessee
Total number of states	13 states	_____ states	26 states
Names of states	Alabama, Georgia, Illinois, Indiana, Kentucky, Michigan, Mississippi, New York, North Carolina, Ohio, Tennessee, Virginia, West Virginia	_____ _____ _____ _____ _____	Alabama, Arizona, Arkansas, Colorado, Georgia, Illinois, Indiana, owa, Kansas, Kentucky, Louisiana, Maryland, Michigan, Mississippi, Missouri, Nebraska, New York, North Carolina, North Dakcta, Oklahoma, South Carolina, South Dakcta, Tennessee, Texas, Virginia, Wisconsin
Regions	Midwest, Northeast, South	_____ _____	Midwest, Northeast, South, West

Continued on the next page.

	Event 1	Event 2	Event 3
Date	April 3–4, 1974	May 3, 1999	May 1–10, 2003
F-scale	F2 = 34 F3 = 33 F4 = 23 F5 = 7	F2 = 7 F3 = 6 F4 = 3 F5 = 1	F2 = ____ F3 = ____ F4 = ____ F5 = ____
Total number of strong and violent tornadoes	Strong = 67 Violent = 30	Strong = 13 Violent = 4	Strong = ______ Violent = ______
Total number of injuries	________ injuries	825 injuries	637 injuries
Total number of deaths	________ deaths	46 deaths	40+ deaths
Total damage in dollars	>600 million	1.6 billion	3.4 billion

Module 4, Lesson 2

Analyzing historical tornadoes

Handout: Tornado F-Scale

Tetsuya T. Fujita, also known as "Mr. Tornado," invented the Fujita Scale (F-scale) at the University of Chicago in the early 1970s. The F-scale describes different levels of damage caused by tornadoes based on their estimated wind speeds. The scale goes from light damage (F-0) to incredible damage (F-5).

F-scale	Wind speed (miles per hour)	Damage
F0	40–72 mph	Light damage—Some damage to chimneys; breaks twigs and tree branches; pushes over some trees; damages signboards; breaks some windows
F1	73–112 mph	Moderate damage—Peels surfaces off roofs; mobile homes pushed off foundations; moving autos pushed off the roads; trees snapped or broken
F2	113–157 mph	Considerable damage—Roofs torn off houses; mobile homes destroyed; some houses lifted and moved; railroad cars pushed over; large trees snapped or pulled out; light objects take flight
F3	158–206 mph	Severe damage—Roofs and walls torn off houses; whole trains turned over; most forest trees pulled out; heavy cars lifted off the ground and thrown; pavement blown off roads
F4	207–260 mph	Devastating damage—Houses flattened; weak buildings blown some distance; cars thrown and destroyed; large objects take flight; forest trees pulled out and blown away
F5	261–318 mph	Incredible damage—Strong houses lifted and blown away; objects the size of autos fly through the air; bark blown off trees; incredible things happen

Data sources

Module 1 - \OurWorld1\Module 1\data sources include:

\Cntry07 shapefile, from ESRI Data & Maps 2007, courtesy of ArcWorld Supplement.

\continent shapefile, from ESRI Data & Maps 2006, courtesy of ArcWorld Supplement.

\geogrid shapefile, from ESRI Data & Maps 2006, courtesy of ESRI.

\magellan2 shapefile courtesy of the authors.

\minus45plus135 shapefile, from ESRI Data & Maps 2006, courtesy of ESRI.

\Moluccas shapefile, from ESRI Data & Maps 2007, courtesy of ArcWorld Supplement.

\ocean_names shapefile, courtesy of the authors.

\ocean_names2 shapefile, courtesy of the authors.

\Rivers shapefile, from ESRI Data & Maps 2006, courtesy of ArcWorld.

\seville shapefile, from ESRI Data & Maps 2006, courtesy of ArcWorld.

\Stops2 shapefile, courtesy of the authors.

\StudyArea1 shapefile, courtesy of the authors.

\StudyArea2b shapefile, courtesy of the authors.

\StudyArea2c shapefile, courtesy of the authors.

\StudyArea3 shapefile, courtesy of the authors.

\StudyArea_df shapefile, courtesy of the authors.

\world30 shapefile, from ESRI Data & Maps 2006, courtesy of ESRI.

Module 2 - \OurWorld1\Module 1\Data\Pictures sources include:

\Elephant.jpg, courtesy of PhotoDisc/PhotoDisc- Nature, Wildlife and the Environment 2.

\Flamingo.jpg, courtesy of PhotoDisc/PhotoDisc- Nature, Wildlife and the Environment 2.

\Giraffe.jpg, courtesy of PhotoDisc/PhotoDisc- Nature, Wildlife and the Environment 2.

\Gorilla.jpg, courtesy of PhotoDisc/PhotoDisc- Nature, Wildlife and the Environment 2.

\Hippopotamus.jpg, courtesy of PhotoDisc/PhotoDisc- Nature, Wildlife, and the Environment 2.

\Lion.jpg, courtesy of PhotoDisc/PhotoDisc- Nature, Wildlife and the Environment 2

\Panda.jpg, courtesy of PhotoDisc/PhotoDisc- Nature, Wildlife and the Environment 1.

\Tiger.jpg, courtesy of PhotoDisc/PhotoDisc- Nature, Wildlife and the Environment 2.

\Zebra.jpg, courtesy of PhotoDisc/PhotoDisc- Nature, Wildlife and the Environment 1.

Module 2 - \OurWorld1\Module 2\data sources include:

\AnimalRanges shapefile, courtesy of ArcAtlas.

\Biomes shapefile, courtesy of World Wildlife Fund.

\Boudary shapefile, courtesy of authors.

\Buildings shapefile, courtesy of authors.

\Cntry07 shapefile, from ESRI Data & Maps 2007, courtesy of ArcWorld Supplement.

\Continent shapefile, from ESRI Data & Maps 2006, courtesy of ArcWorld Supplement.

\Geogrid shapefile, from ESRI Data & Maps 2006, courtesy of ESRI.

\Habitats shapefile, courtesy of authors.

\Status shapefile, courtesy of World Wildlife Fund.

\Walkway shapefile, courtesy of authors.

\World30 shapefile, from ESRI Data & Maps 2006, courtesy of ESRI.

\ZooAnimals shapefile, courtesy of the authors.

Module 3 - \OurWorld1\Module 3\data sources include:

\cities shapefile, from ESRI Data & Maps 2006, courtesy of US Census.

\comp_usage shapefile, courtesy of US Census.

\lakes shapefile, from ESRI Data & Maps 2006, courtesy of ArcWorld.

\NERivers shapefile, from ESRI Data & Maps 2006, courtesy of ArcWorld.

\poly5000 shapefile, from AEJEE, courtesy of USGS and Digital Chart of the World.

\prism0p020_Dis2 shapefile, courtesy of National Atlas.

\Rivers shapefile, from ESRI Data & Maps 2006, courtesy of ArcWorld.

\sm_cntry shapefile, from ESRI Data & Maps 2006.

\StateLevel1 shapefile, from ESRI Data & Maps 2006, courtesy of ArcUSA, US Census, and ESRI.

\StateLevel3 shapefile, from ESRI Data & Maps 2006, courtesy of ArcUSA, US Census, and ESRI.

\states shapefile, from ESRI Data & Maps 2006, courtesy of ArcUSA, US Census, and ESRI.

\states_1 shapefile, from ESRI Data & Maps 2006, courtesy of ArcUSA, US Census, and ESRI.

\US48SHD2.jpg, from AEJEE, courtesy of USGS.

\US48SHD2.prj, from AEJEE, courtesy of USGS.

\world30 shapefile, from ESRI Data & Maps 2006, courtesy of ESRI.

\WWF_terra_Clip shapefile, courtesy of World Wildlife Fund.

Module 4 - \OurWorld1\Module 4\data sources include:

\1974_2003_T shapefile, courtesy of National Atlas.

\April3_4_74 shapefile, courtesy of National Atlas.

\Density_states2, courtesy of National Atlas.

\F3to5_95to04 shapefile, courtesy of National Atlas.

\Fall_T shapefile, courtesy of National Atlas.

\lakes shapefile, from ESRI Data & Maps 2006, courtesy of ArcWorld.

\May1_10_2003 shapefile, courtesy of National Atlas.

\May3_1999 shapefile, courtesy of National Atlas.

\Moore shapefile, data from ESRI Data & Maps 2006, courtesy of US Census, combined with data courtesy of National Atlas.

\Names2 shapefile, courtesy of the authors.

\sm_cntry shapefile, from ESRI Data & Maps 2006.

\Spring_T shapefile, courtesy of National Atlas.

\State_tornadoes shapefile, from ESRI Data & Maps 2006, courtesy of ArcUSA, US Census, and ESRI.

\states_1 shapefile, from ESRI Data & Maps 2006, courtesy of ArcUSA, US Census, and ESRI.

\states_1_Dissolve, from ESRI Data & Maps 2006, courtesy of ArcUSA, US Census, and ESRI.

\Strong_T40 shapefile, courtesy of National Atlas.

\Summer_T shapefile, courtesy of National Atlas.

\Torn9504states, courtesy of National Atlas.

\torn_cities3 shapefile, data from ESRI Data & Maps 2006, courtesy of US Census, combined with data courtesy of National Atlas.

\US48SHD2.jpg, from AEJEE, courtesy of USGS.

\US48SHD2.prj, from AEJEE, courtesy of USGS.

\Violent_T40 shapefile, courtesy of National Atlas.

\Weak2_40 shapefile, courtesy of National Atlas.

\Winter_T shapefile, courtesy of National Atlas.

\world30 shapefile, from ESRI Data & Maps 2006, courtesy of ESRI.

\Xenia shapefile, data from ESRI Data & Maps 2006, courtesy of US Census, combined with data courtesy of National Atlas.

16 9890